Basic Statistics Using SAS® Enterprise Guide®
a Primer

Geoff Der
Brian S. Everitt

The correct bibliographic citation for this manual is as follows: Der, Geoff, and Brian S. Everitt. 2007. *Basic Statistics Using SAS® Enterprise Guide®: A Primer.* Cary, NC: SAS Institute Inc.

Basic Statistics Using SAS® Enterprise Guide®: A Primer

Contents

Preface

SAS Enterprise Guide provides a graphical user interface to SAS. Because it is so much easier to use and quicker to learn than the traditional programming approach, SAS Enterprise Guide makes the power of SAS available to a much wider range of potential users. The aim of this book is to offer further encouragement to users by showing how to conduct a range of statistical analyses within SAS Enterprise Guide. The emphasis is very much on the practical aspects of the analysis. In each case, one or more real data sets are used. The statistical techniques are briefly introduced and their rationale explained. They are then applied using SAS Enterprise Guide, and the output is explained. No SAS programming is needed, only the usual Windows point-and-click operations are used and even typing is kept to a bare minimum. There are also exercises at the end of each chapter to summarize what has been learned. All the data sets and solutions to exercises are available for downloading from this book's companion Web site at support.sas.com/companionsites so that users can work through the examples for themselves. Give it a try!

We would like to thank Julie Platt and the rest of the SAS Press team for their constant help and encouragement during the writing and production of this book.

Geoff Der and Brian S. Everitt
Glasgow and London 2007

x

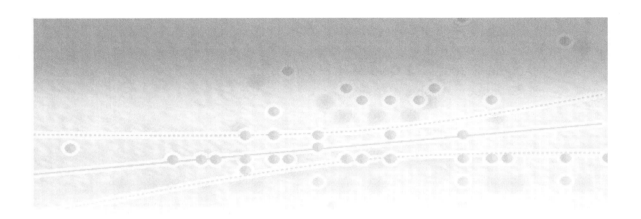

Chapter 1

Introduction to SAS Enterprise Guide

1.1 What Is SAS Enterprise Guide?

SAS is one of the best known and most widely used statistical packages in the world.
Although it actually covers much more than statistical analysis, that is the focus of this
book. Analyses using SAS are conducted by writing a program in the SAS language,
running the program, and inspecting the results. Using SAS requires both a knowledge of
programming concepts in general and of the SAS language in particular. One also needs
to know what to do when things don't go smoothly; i.e., knowing about error messages,
their meanings, and solutions.

SAS Enterprise Guide is a Windows interface to SAS whereby statistical analyses can be
specified and run using normal windowing point-and-click style operations and hence
without the need for programming or any knowledge of the SAS programming language.
As such, SAS Enterprise Guide is ideal for those who wish to use SAS to analyze their
data, but do not have the time, or perhaps inclination, to undertake the considerable
amount of learning involved in the programming approach. For example, those who have
used SAS in the past, but are a bit "rusty" in their programming, may prefer SAS
Enterprise Guide. Then again, those who would like to become proficient SAS
programmers could start with SAS Enterprise Guide and examine the programs it
produces.

It should be born in mind that SAS Enterprise Guide is not an alternative to SAS; rather,
it is an *addition* which allows an alternative way of working. SAS itself needs to be
present or at least available. The need for SAS to be present is because SAS Enterprise
Guide works by translating the user's point-and-click operations into a SAS program.
SAS Enterprise Guide then uses SAS to run that program and captures the output for the
user.

The computer on which SAS runs is referred to as the *SAS Server*. Usually the SAS
Server will be the same computer, referred to as the *Local Computer*, but need not be. We
assume that both SAS and SAS Enterprise Guide will have already been set up. The

examples in this book were produced using SAS Enterprise Guide 4.1 and SAS 9.1 under Windows XP Professional. There are some notable differences between version 4.1 and earlier versions, so we would encourage users of earlier versions to upgrade. Such upgrades are available from your local SAS office.

1.2 Using This Book

We assume readers are familiar with the basic operation of Windows and Windows programs; for example, we will use the terms: click, right-click, double-click, and drag to refer to the usual mouse operations without further comment. The description of how to perform a task within SAS Enterprise Guide will usually begin from one of the main menus and typically comprise a sequence of selections from there. For instance, the **File** menu contains the usual **Open** option within it, the use of which leads to a submenu of the kinds of things that can be opened, one of which is **Data**. We abbreviate this sequence to **File≻Open≻Data**. When it seems natural we may extend the sequence to options within the windows that open as a result of the menu selection. Thus, the window that opens following the above sequence (shown in Display 1.5) has two options: **Local Computer** and **SAS Servers**, so the sequence might be extended to **File≻Open≻Data≻Local Computer**. We use the bold, sans-serif font both to distinguish text that appears on screen and forms part of the operation of SAS Enterprise Guide and to distinguish the names of data sets and variables from ordinary text.

Many of our instructions assume that the downloadable files and data sets that accompany this book have been placed in the directory **c:\saseg** and its subdirectories **data** and **sasdata**. If they have been placed elsewhere, the instructions will need to be amended accordingly.

This introductory chapter includes numerous screenshots, whereas subsequent chapters use fewer and rely on the more concise sequences of instructions. It is assumed that the reader will have downloaded the data and will be able to follow the instructions on screen.

In the production of this book, we have altered several settings from their defaults. Readers may wish to use the same settings for comparability between the results shown here and their own results and they can do this, by first make sure settings are at their defaults, by selecting **Tools≻Options≻Reset All**.

Then make the follow changes:

- **Tools≻Options≻Results≻General,** select **RTF** and deselect **HTML**. Click **OK**.
- **Tools≻Options≻Results≻RTF**, select **Theme** as the **Style**. Click **OK**.

- **Tools≻Options≻Tasks≻Tasks General**, delete the **Default footnote text for task output,** and deselect **Include SAS procedure title in results**. Click **OK**.

- **Tools≻Options≻Query**, select the option to **Automatically add columns from input tables to result set of query**. Click **OK**.

1.3 The SAS Enterprise Guide Interface

When SAS Enterprise Guide starts, it first attempts to connect to SAS servers that it knows about. In most cases, connecting to SAS servers simply means that it finds that SAS is installed on the same computer. SAS Enterprise Guide then offers to open one of the projects that have recently been opened or to create a new project as shown in Display 1.1.

Display 1.1 Welcome Screen

1.3.1 **SAS Enterprise Guide Projects**

A *project* is the way in which SAS Enterprise Guide stores statistical analyses and their results: it records which data sets were used, what analyses were run, and what the results were. It can also record the user's own notes on what they did and why. In the same way that a word processor loads and saves documents, so SAS Enterprise Guide does with projects. Thus, a project is a piece of statistical analysis in the same way that a document is a piece of writing. In terms of scope, a project might be the user's approach to answering one particular question of interest. It should not be so large or diffuse that it becomes difficult to manage.

1.3.2 **The User Interface**

The default user interface for SAS Enterprise Guide 4.1 is shown in Display 1.2.

Display 1.2 SAS Enterprise Guide User Interface

The most familiar elements of the interface are the menu bar and toolbar at the top of the window. There are four windows open and visible:

❶ the Project Explorer window

❷ the Project Designer window

❸ the Task Status window

❹ the Task List window

Moving the cursor over the task list causes the task list to scroll to the right.

For the vast majority of the examples in this book, we use only the menus and the Project Designer window. In this way the reader can safely ignore other elements of the interface, or even close them. We give a brief description of them, for completeness sake.

Toolbar and Task List	offer alternative, sometimes quicker, ways to access features of SAS Enterprise Guide.
Task Status window	shows what is happening while SAS Enterprise Guide is using SAS to run a program.
Project Explorer window	offers an alternative view of the project to that presented in the Project Designer window. It tends to show more detail, which can be useful in some cases.

1.3.3 The Process Flow

Within the Project Designer window, we can see an element labeled **Process Flow**, which is another concept central to SAS Enterprise Guide. Essentially, a process flow is a diagram consisting of icons that represent data sets, tasks, and outputs with arrows joining them to indicate how they relate to each other. The general term *tasks* includes not only statistical analyses but data manipulation.

We will begin with some examples of process flow diagrams to give an overview before describing the individual elements in more detail. An example of a Project Designer window is shown in Display 1.3.

Display 1.3 An Example of a Project Designer Window

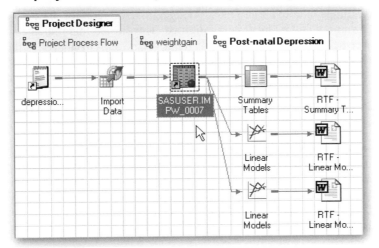

The first thing to note about this example is that the Project Designer window actually contains three process flows, identified by tabs at the top of the window:

- Project Process Flow (the default name)
- weightgain
- Post-natal Depression

To make a process flow active and bring it to the front, click on the tab. In this case, the Post-natal Depression process flow is the active one, and the title on the tab is bold to indicate that this is the case.

The first three icons in Display 1.3 represent the process of importing some data into a SAS data set. The Import Data task has as its input a raw data file, depressionIQ (**depressio...**), and as its output a SAS data set. The full name of the raw data file is not visible in the process flow; if the cursor is held over the icon, a window pops up with more details, including the full name, path, and location (i.e., which computer it is on). The SAS data set has been automatically given the somewhat arbitrary name **SASUSER.IMPW_0007**. The relationship of a task to its input and output is represented primarily by the arrows, but also by the ordering from left to right—input to the left of the task and output to the right of the task.

On the right-hand side of the process flow diagram, we can see that the SAS data set is used as input to three tasks: a **Summary Tables** task and two **Linear Models** tasks. The output from each task is an RTF (rich text format) document containing the results. RTF is one of the formats that can be chosen for output and is one particularly suited for reading into a word processor.

1.3.4 **The Active Data Set**

Two important things to note about Display 1.3 are that the icon for the SAS data set has a dashed line around it and its label is highlighted. The dashed line indicates that the SAS data set has been selected (clicked), and this makes it the active data set. If there are multiple data sets in a project, any tasks selected from the menus will apply to the active data set. It is therefore important to be aware of which data set is active and of how to make a data set active. Each type of object and task in the process flow has its own icon, and a SAS data set can be recognized by the icon (the grid with the red ball in the bottom right corner).

A second example, shown in Display 1.4, contains four SAS data sets. The first data set results from importing some raw data from a file named **LENGTHS**, and the other data sets are derived from it. Generating other data sets is a common situation, where there is an original data set and one or more different versions arise from some modification of the original data. The **feet** data set is the active data set, so any analysis chosen from the menus would apply to that data set.

Display 1.4 A Process Flow Containing Multiple SAS Data Sets

Any of the icons in a process flow diagram can be opened by double-clicking them or right-clicking, and selecting **Open**. For a file, data set, or output, the contents can then be examined, printed, or copied. For a task, the settings can be examined, changed if required, and the task re-run. When a task is re-run, there is the option to replace the output from the previous run or generate new output, keeping the previous version. If the Replace option is taken, a new task icon and output icon will appear in the process flow.

1.4 Creating a Project

The first step in a project is adding the data. In order to be analyzed, data must be in the form of a SAS data set. Data in other formats will need to be converted or imported into a SAS data set. In many cases, the conversion or importation will have already been done.

1.4.1 Opening a SAS Data Set

To add a SAS data set to a project, select **File≻Open≻Data**. A window like that shown in Display 1.5 will then appear, prompting a location from which to open the data. **Local Computer** is the user's own computer where SAS Enterprise Guide is being used. **Local Computer** would also be the location for data stored on a network file server mapped to a local drive letter. For example, if the user had data stored on a network drive N: that would also count as stored on the local computer. The alternative, **SAS Servers**, refers to remote computers that have SAS installed and hold SAS data sets. All of the examples in this book use data stored on the local C: drive.

Display 1.5 Data Location Pop Up Window

Having selected **Local Computer** or a **SAS Servers**, browse to the location of the SAS data set, select it, and click **Open**. In our examples, SAS data sets are stored in the directory **c:\saseg\sasdata**. SAS data sets created with version 7 of SAS or a later version have the extension .sas7bdat. Data sets created by earlier versions of SAS are most likely to have the extension .sd2. The SAS data set **water.sas7bdat** contains measures of water hardness and mortality rates for 61 towns in England and Wales. Open that data set and the contents of the data set can then be viewed on screen as shown in Display 1.6.

Display 1.6 The Water Data Set Opened

	Town	Mortal	Hardness	location
	Project Designer	water (read-only)		
1	Bath	1247	105	south
2	Birkenhead	1668	17	north
3	Birmingham	1466	5	south
4	Blackburn	1800	14	north
5	Blackpool	1609	18	north
6	Bolton	1558	10	north
7	Bootle	1807	15	north
8	Bournemouth	1299	78	south
9	Bradford	1637	10	north
10	Brighton	1359	84	south
11	Bristol	1392	73	south
12	Burnley	1755	12	north

Closing the data set, we see that a SAS data set icon, labeled **water**, has been added to the process flow.

1.4.2 Importing Data

If the data to be analyzed are not already available as a SAS data set, they need to be imported into one, using the Import Data task. We begin with examples of importing raw data files, which are also referred to as text files or ASCII files. Such files contain only the printable characters plus spaces, tabs, and end-of-line characters. The files produced by database programs and spreadsheets are not normally in this format, although the programs usually have an export facility to create raw data files.

The data in a raw data file may be fixed width or delimited. With fixed-width data, the values for each variable are in prespecified columns. With delimited data, the data values are separated by a special character—usually a space, tab, or comma. Tab-separated files and comma-separated files are very common formats. Comma-separated data are sometimes referred to as *comma-separated values* and given the extension .csv. Delimited files may also contain the names of the variables, usually as the first line of the file, with the names separated by the same delimiter as the data values.

There are examples of importing both tab- and comma-delimited data, with and without the variable names, in later chapters (see the index). Here, we illustrate the use of the Import Data task with fixed-width data. The water.dat file contains a slightly different version of the data already available in the SAS data set of the same name. To import them, select **File≻Import Data**.

The Import Data task, as with most tasks, consists of a number of panes, each of which allows a set of options to be specified. The initial view is shown in Display 1.7.

Display 1.7 Import Data Task Opening Screen

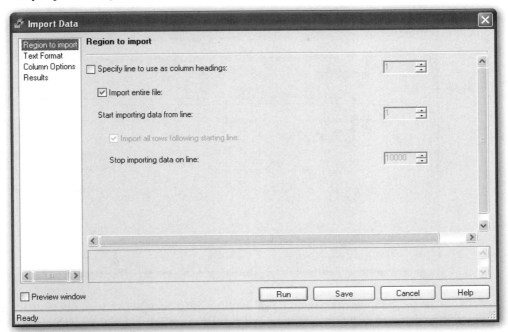

The first pane, **Region to import**, is displayed. Other panes, listed in the left side of the window, are: **Text Format**, **Column Options**, and **Results**. In the **Region to import** pane, **Import entire file** is the default. The option to **Specify line to use as column headings** is for delimited files where the variable names are included in the file, usually in line 1. Hence, 1 is the default value if the option is selected. The **Text Format** pane allows the format to be specified as **Fixed Width** or **Delimited** and, if delimited, what delimiter is used. The default is comma-delimited. Display 1.8 shows the result of selecting **Fixed Width** format with this data file.

Display 1.8 Text Format Pane for Water Data

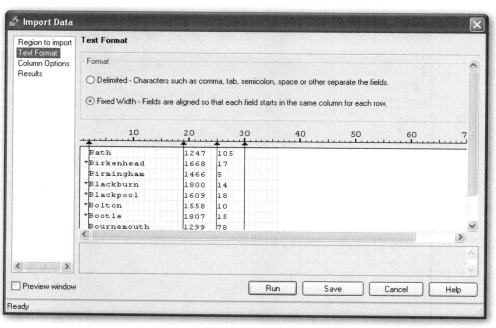

The pane shows the beginning of the file with a ruler above to indicate which columns the data values are in. Clicking on the ruler specifies where the data fields begin and end. We have put the separators at columns 2, 19, 25, and 30. The Column Options pane is shown in Display 1.9.

Display 1.9 Column Options Pane for Water Data

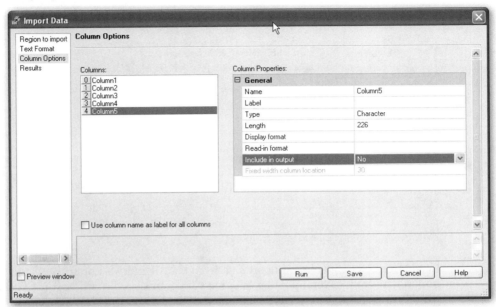

We see first that five rather than four columns have been defined. Column 5 is the blank remainder of the line after the final delimiter, so we have set the **Include in output** option to **No**. In the pane shown in Display 1.9, we can also give the variables (or columns) more meaningful names. Select **Name** under Column Properties and type a new name. Rename columns 1 to 4 as **flag**, **town**, **Mortality**, and **hardness**, respectively. (We deselected the option to **Use column names as label for all columns** to avoid having to retype these labels as well.)

We also check that other properties of the columns have been correctly assigned. In fact, **Mortality** and **hardness** have been treated as character variables when they should be numeric, but we can change the variable type using the **Type** option under Column Properties.

The final **Results** pane allows the SAS data set being created to be renamed and stored in a particular location. In this case, we leave the default settings and run the task. Display 1.10 shows the results, which are similar to the results shown previously in Display 1.6. The data set has been given an arbitrary name, **SASUSER.IMPW_000A**. At this point, we should scroll through the data to make sure it has all been imported correctly. Having done that, we would close the **water** data set as its contents are in front of the process flow. We could click on the process flow tab (labeled **Project Designer**) to bring it to the front, but it keeps the workspace tidier if we close data sets and output after we have viewed them.

Display 1.10 Imported Version of Water Data

	flag	town	Mortality	hardness
1		Bath	1247	105
2	*	Birkenhead	1668	17
3		Birmingham	1466	5
4	*	Blackburn	1800	14
5	*	Blackpool	1609	18
6	*	Bolton	1558	10
7	*	Bootle	1807	15
8		Bournemouth	1299	78
9	*	Bradford	1637	10
10		Brighton	1359	84
11		Bristol	1392	73
12	*	Burnley	1755	12

In addition to being able to import data from text files, SAS Enterprise Guide can also import data from several popular Windows programs such as Microsoft Excel and Microsoft Access. As a simple example, the file **c:\saseg\data\usair.xls** contains a Microsoft Excel workbook with some data on air pollution in the USA. The data are described more fully in Chapter 6 (Exercise 6.4) but need not concern us here. To import the data:

1. Select **File≻Import Data≻Local Computer**.

2. Browse to **c:\saseg\data**.

3. Select **usair.xls** and **Open**. Because the file contains more than one worksheet and only one can be imported at a time, a window like that in Display 1.11 pops up to select the worksheet to use.

4. Select **USAIR** and then **Open**. The worksheet contains the variable names in the first row. SAS Enterprise Guide has recognized this and set the options under **Region to import** and **Column Options** appropriately, so no changes are needed.

5. Run the task. It is worth noting that the ease of importing the data is due to the fact that the spreadsheet contains only the variable names and the data values. It would be simpler again if the file contained only a single worksheet.

Importing a data table from an Access database would be very similar. It may also be possible to open or import data (**File**➤**Open**➤**Data** or **File**➤**Import Data**) from other proprietary databases, if the appropriate component of SAS (a module of SAS/ACCESS) has been licensed for the computer running SAS.

Display 1.11 Table Selection Window

1.5 Modifying Data

After adding data to a project, it may be necessary to modify the data before it is ready to be analyzed. The Filter and Query task can be used to modify a SAS data set in a variety of ways.

1.5.1 Modifying Variables: Using Queries

We begin with an example of creating a new variable from an existing variable. One common reason for creating a new variable is when a transform of an existing variable is considered necessary. The **hardness** variable in the **water** data set is somewhat skewed, so a log transformation might be appropriate.

1. Click on the **water** data set to make it active. There are two icons in the process flow both named water. The SAS data set that we wish to use is distinguished by its icon—the text file of the same name has a notepad icon. They can also be distinguished by holding the cursor over them, which reveals additional details of each.

2. Select the SAS data set.

3. Select **Data➤Filter and Query**. The opening screen should look like Display 1.12.

Display 1.12 Query Builder Window

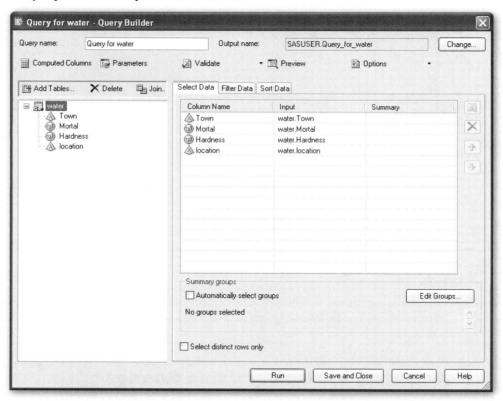

The four variables in the input data set also appear in the **Select Data** pane because we have set the option to **Automatically add columns from input tables to result set of query** under **Tools➤Options➤Query**. Otherwise, variables from the input data set would need to be dragged across. It is worth noting in passing that the variables have icons that indicate whether they are character or numeric.

4. To create a new variable, select **Computed Columns≻New≻Build Expression**. This brings up the **Advanced Expression Editor** window as shown in Display 1.13.

Display 1.13 Advanced Expression Editor

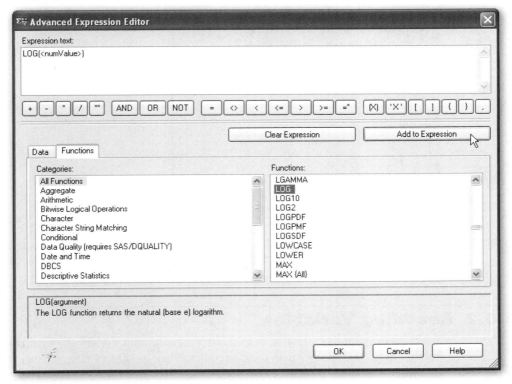

The expression text specifies how the new variable is to be calculated. It can either be typed into the pane or constructed using the buttons and menus. Selecting the **Functions** tab shows a list of function categories with **All Functions** as the default. The right hand pane shows the functions by name, with a brief description of the highlighted function below.

5. Scroll down this list, click on **LOG** and **Add to Expression**. **LOG(<numValue>)** appears in the expression text. The **<numValue>** part indicates that the log function takes a numeric argument.

6. Because we want the log of the **hardness** variable, replace **<numValue>** with **hardness** either by simply typing **hardness** in or by using the Data tab. If the Data tab is used, the variable name will be prefixed with the name of the data set.

7. Clicking **OK** returns us to the Computed Columns window as shown in Display 1.14. The new variable is simply called **Calculation1**, by default, but can be renamed by selecting it, clicking **Rename**, and typing in a more meaningful name, such as **loghardness**.

Display 1.14 Computed Columns Window

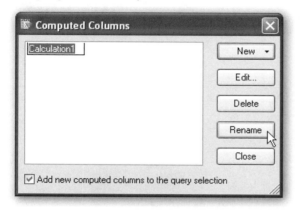

Running the task adds an icon for the query and a new SAS data set to the process flow. The new data set contains the **loghardness** variable in addition to the original four variables.

1.5.2 Recoding Variables

Another common modification is to classify a continuous variable like **hardness** into a number of groups. Rather than create another Filter and Query task, we can re-open the existing one and add to that.

1. Open the task by double-clicking on its icon, or by **right-click≻Open**.

2. Select **Computed Columns≻New≻Recode a column**.

3. Select **hardness** and **Continue**. The **Recode Column** window opens.

4. Click on the **Add** button.

5. Select the **Replace a range** tab.

6. Use these to replace the ranges 0–15 with 1, 16–60 with 2, and 61–138 with 3. The actual values of **hardness** contained in the data are available to view via the drop-down boxes for the start and end of the ranges. The **Recode Column** window

should now look like Display 1.15. Change the **New column name** to **hardness3groups** as shown.

7. Click **OK**, **Close**, and **Run**.

8. Reply **Yes** to **Would you like to replace the results from the previous run?** The **Recode Column** option within the Filter and Query task can also be used to reduce the number of categories a categorical variable has, for instance when combining categories which have too few members in. Such recoding can be done with both numeric and character variables.

Including multiple data modifications in the one Filter and Query task helps to keep the process flow diagrams simple and clear.

Display 1.15 Recode Column Window

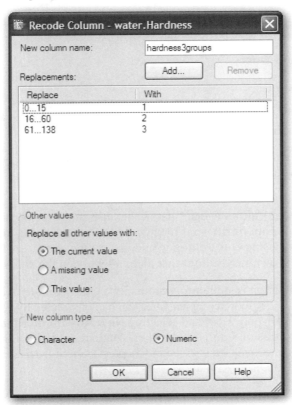

To modify the value of a variable for some observations and not others, or to make different modifications for different groups of observations, use the Advanced Expression Editor to build a query with a conditional function. A simple example is given in Chapter 2, Section 2.3.1.

1.5.3 Splitting Data Sets: Using Filters

So far we have looked at using the Filter and Query task to create and modify the values of variables and we used queries for the purpose. We now turn to the use of filters to produce subsets of the observations in a data set. We might want to form a subset of the observations in order to discard observations that have errors, or because we wish to focus our analysis on one particular group of observations. Take the **water** data set as an example where we want to look only at the northerly towns. Normally we would want to include the newly derived variables, and so we would use the data set calculated with the query described above.

1. Click on the **water** data set to make it the active data set.

2. Select **Data➢Filter and Query**.

3. Click on the **Filter Data** tab.

4. **Location** is the variable we want to filter on, so we drag and drop that into the **Filter Data** pane. The **Edit Filter** window pops up.

5. The value of location that we want to select is **north**. We could simply type that into the value box, but it would be safer to use the drop-down button and select **Get Values**.

The reason for preferring **Get Values** is that filters which use character variables are case sensitive: **North** is different from **north**, so if both occurred in the data set, the filter would need to include both. Using **Get Values** would give us the correct spelling and case as well as alerting us to any misspellings that there might be in the data set.

In our example here, the situation is straightforward and the Query Builder window should look like Display 1.16. A more complex filter can be constructed by clicking the new filter button (circled in Display 1.16) and selecting **New Advanced Filter**, which brings up the Advanced Expression Editor seen earlier. Another example of using filters to split the data set for separate analyses is given in Chapter 2, Section 2.2.2, and the process flow is reproduced in Display 1.4 above.

Display 1.16 Query Builder Window Filtering the Water Data Set

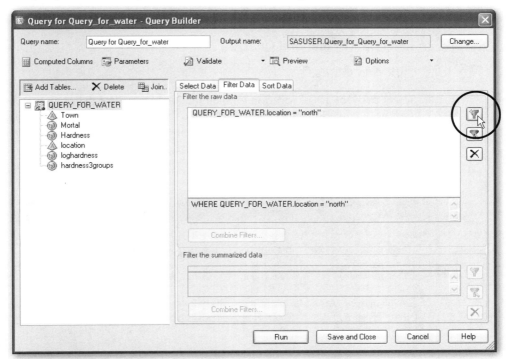

1.5.4 Concatenating and Merging Data Sets: Appends and Joins

Where two or more data sets contain the same variables (or mostly the same) but different observations, they can be combined into a single data set using **Data▷Append Table** and specifying the table(s) to be concatenated with the active data set. Concatenation is essentially the converse of the process of splitting data sets described above.

Where two data sets contain mostly the same observations but different variables, they can be combined to create a data set with all the variables using a join. Joins are yet another function of the Filter and Query task. We will illustrate a join again using the **water** data set. The original **water** data set has a variable, **location**, with values **north** and **south**. The version imported from the raw data has a variable, **flag**, where the value

'*' indicates the more northerly towns. To check that the two variables do in fact correspond, we will merge the data sets to produce one that has both variables.

1. Make the imported data set the active data set.

2. Select **Data▷Filter and Query**.

3. Click **Add tables**.

4. Select **project** as the location to open the data from. The list of similarly named data sets shown in Display 1.17 illustrates the potential value of giving output data sets explicit and more meaningful names. In this instance, the one simply labeled **water** is the one we need.

Display 1.17 List of Project Data Sets

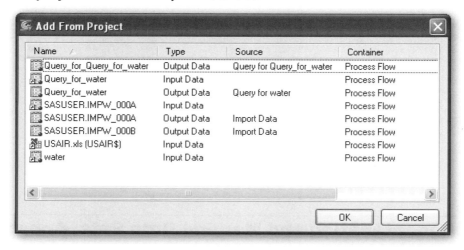

5. Select the **water** data set.

6. Click **OK**. A Query Builder window like that shown in Display 1.18 opens.

Display 1.18 Query Builder Window for Join of Two Versions of the Water
Data Set

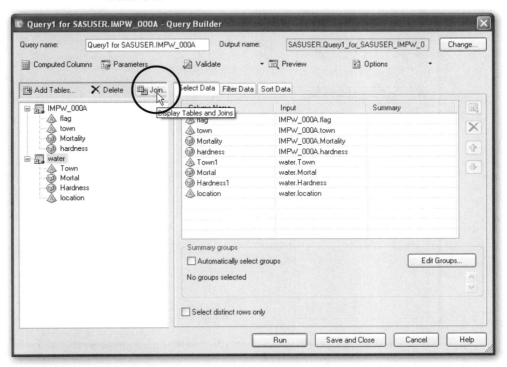

All the variables from the **water** data set have been added and, where they had the same
name, the names have been suffixed with a 1 to make them distinct.

7. Click on **Join**. The join is displayed, as in Display 1.19, and can be
modified if necessary.

Display 1.19 Join of Two Versions of the Water Data Set

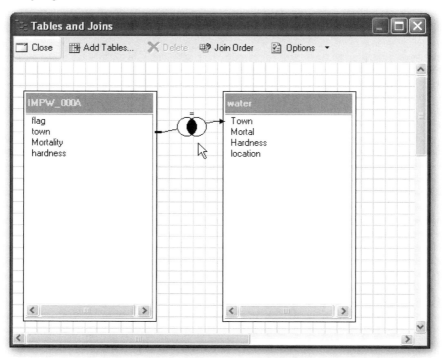

The program has recognized that both data sets contain the variable **town**, which uniquely identifies each observation and can therefore be used to match them. The Venn diagram in the arrow connecting them shows that an inner join will be used. Right-clicking on the Venn diagram and selecting **Modify Join** lists the different types of joins and explains them. A choice will need to be made if the two data sets contain different observations. Here, the two data sets contain the same observations, so the type of join makes no difference.

8. Close the **Tables and Joins** window.

9. Use the buttons on the right of the **Select Data** pane to delete **Town1**, **Mortal**, and **Hardness1**, and to move **flag** next to **location**.

10. Run the query.

11. Sort the resulting data set by location (**Data➢Sort Data** and **Sort by location**). Scrolling down the results confirms that **flag** and **location** do indeed correspond.

The process flow should now resemble Display 1.20. It is beginning to look a bit confusing. Several tasks and data sets have similar names (beginning with "Query") which do not give much idea of their purpose or contents.

Display 1.20 Process Flow with Default Names

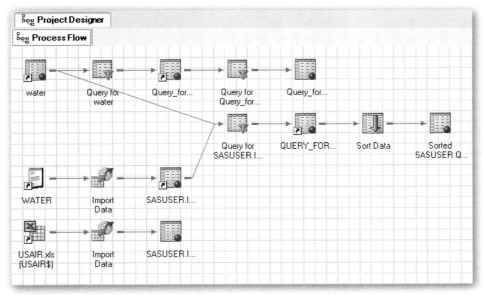

Some of the tasks and data sets could be renamed (**right-click➤Rename**) to make this clearer. Display 1.21 shows an example.

Display 1.21 Process Flow with Renamed Tasks and Data Sets

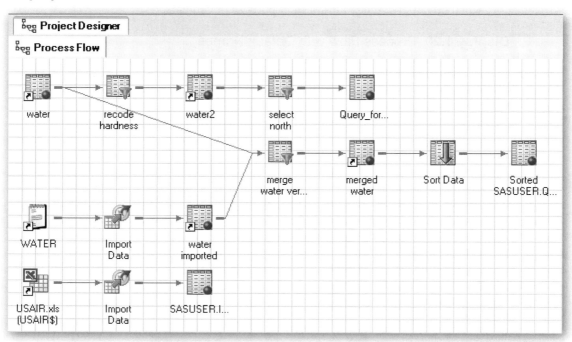

1.5.5 Names of Data Sets and Variables in SAS and SAS Enterprise Guide

Renaming some data sets and tasks in the process flow, as we did for Display 1.21, actually changed their *labels* rather than their *names*. Data sets, variables, and tasks all have labels as well as names, but there are different rules for creating names and labels.

The SAS rules for names of variables and data sets:

- Names are limited to 32 characters or less.
- Names start with a letter or underscore (_) and include only letters, numbers, and underscores. Names should not contain spaces.

Although SAS Enterprise Guide has more flexibility in its naming, we recommend keeping to the SAS rules for variables and data sets.

Labels, in contrast, can contain spaces and other characters and can be up to 256 characters long. However, when there is any doubt about which is being changed, it would be safer to leave spaces out and keep to the rules for SAS names.

1.5.6 Storing SAS Data Sets: Libraries

The SAS data sets created so far have been left with default names and locations. Some data set labels were altered to make the process flow easier to read. In most cases, it is not necessary to alter names and locations. When you want to control where project data sets are stored, use *libraries*. Essentially, a library is a folder where SAS data sets are stored. Rather than refer to the folder explicitly, the folder is assigned an alias: the library name. For example, the data sets created by the Import Data task were automatically given names like **SASUSER.IMPW_xxxx**. The part of the name before the period, **SASUSER**, is the library name and is an alias for **c:\My SAS Files\9.1** on our system (it may vary depending on how SAS Enterprise Guide was set up). To store data sets in a particular folder:

1. Assign a library name for that folder using the **Assign Library** wizard (**Tools➤Assign Library**).

2. Type in a name, which should follow the rules for data set names but be eight characters or less; e.g., **ch1**.

3. Add a description if required.

4. When prompted, browse to the path of the folder; e.g., **c:\saseg\libraries\ch1**.

5. Continue through the wizard accepting defaults and an **Assign Library** icon should be added to the process flow.

   This needs to be run before the library can be used in the project, so it is best to set up the libraries at the beginning of the project. Having set up the library, any data set that is given a name beginning with **ch1.**, such as **ch1.water**, will be stored in the folder **c:\saseg\libraries\ch1**.

All SAS data sets are stored in a library. If a data set name is not prefixed with a library name, it has the implicit library name of WORK which, like SASUSER, is one of the libraries assigned automatically by SAS Enterprise Guide. However, WORK is a temporary library which means that data sets stored in it will be deleted and removed from the project when SAS Enterprise Guide is closed, although the option to move the data sets to another library is offered at that point.

1.6 Statistical Analysis Tasks

Once data in a SAS data set have been added to a project, whether directly or by importing raw data, the analysis can begin. Individual tasks are described in detail in subsequent chapters. Here, we describe some general features of the analysis tasks.

One point to bear in mind is that not all tasks that might be considered as *analysis* are under the **Analyze** menu. Several are accessed from the **Describe** menu, and some of the tasks under the **Data** menu could form part of an analysis.

A typical analysis task consists of a number of panes, each of which allows some aspect of the analysis or set of options to be specified. We begin by looking at an example taken from Chapter 5. The process flow diagram is shown in Display 1.3. Opening the first of the Linear Models tasks gives the screen shown in Display 1.22.

Display 1.22 Linear Models Task Opening Window

The panes are listed down the left: **Task Roles, Model, Model Options**, etc.

The **Task Roles** pane, which is selected, is where the variables that are to be used in the analysis are selected and their roles in the analysis specified. The available variables are listed in the central section, and they can be dragged from there to the specific roles in the right-hand section. The available roles vary depending on the task, but some of the most common are included here:

- The **Dependent variable** is the response variable, the one whose values we are modeling. The numeric icon to the left indicates that only numeric variables can be assigned this role and (**Limit: 1**) to the right indicates that only one response variable can be included in the model. The variable **ChildIQ** has been assigned this role.

- **Quantitative variables** are also numeric. The dashed line around it shows that it has been selected (clicked on) and a description of the role appears in the box below, explaining that these are continuous explanatory variables. There are no variables assigned to this role.

- **Classification variables** are discrete explanatory variables. They can be numeric or character. If they are numeric, classification variables will tend to have relatively few distinct values. **Pa_history** and **Mo_depression** are both assigned this role.

- **Group analysis by** variables are also discrete, numeric, or character—variables which define groups in the data. When a variable is assigned this role, the analysis is repeated for each group defined by the variable. For example, if a variable, **sex**, with values **male** and **female** was assigned this role, the analysis would be repeated for males and females separately. We saw earlier how to use Filter and Query to split or subset a data set. If the reason for doing this is to apply the same analysis to separate groups of observations, then using **Group Analysis by** with a suitable variable could be both simpler and more efficient.

- **Frequency count** variables are used with grouped data, where each observation represents a number of individuals. The **frequency count** variable is the one which specifies how many individuals the observation pertains to. The most common use is in analysing tabulated data. Examples are given in Chapter 3, Sections 3.4.3 and 3.4.4.

- The **relative weight** role is for weighted analysis.

Task panes like **Model, Model Options**, and **Advanced Options**, as their names imply, specify what model is to be fitted and how. They will be dealt with in detail in later chapters as they arise.

Many analysis tasks also produce plots of data values, predicted values, residuals, etc., each of which may be specified in the **Plots** pane(s).

1.7 Graphs

SAS Enterprise Guide also makes the powerful graphics facilities of SAS much easier to use. Some of these graphic facilities are available within analysis tasks and others are accessed from the **Graph** menu. A wide range of plots and charts are described in later chapters. Rather than describe the graph tasks here, the interested reader is referred to the index.

One point to note, however, is that the graphs produced are dependent both on the format of the results and the graph format. Both formats are specified under **Tools➤Options➤Results➤Results General** and **Tools➤Options➤Results➤Graph**. One major difference is that, when the output format is RTF, the graphs are included in the same file as the textual output and tables; when HTML output is chosen, each graph appears in a separate file with its own icon in the process flow.

1.8 Running Parts of the Process Flow

So far, we have described running individual tasks. It is also possible to run a branch of the process flow or the whole process flow. If we right-click on any task within a process flow, we will have the option to run that task or to run the branch from that task. The branch is everything to the right of the task which is directly or indirectly connected to it by the arrows. To run the whole process flow, right-click on its tab and select **Run**.

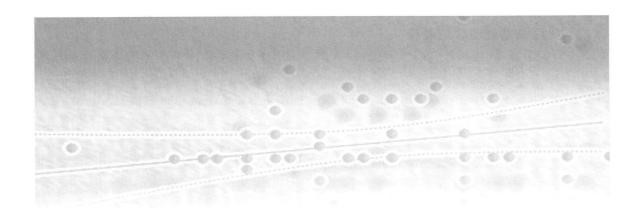

Chapter 2

Data Description and Simple Inference

2.1 Introduction

In this chapter, we will describe how to get informative numerical summaries of data and graphs which allow us to assess various properties of the data. In addition, we will show how to test whether different populations have the same mean value. The statistical topics covered are:

- Summary statistics such as means and variances
- Graphs such as histograms and box-plots
- Student's *t*-test

2.2 Example: Guessing the Width of a Room: Analysis of Room Width Guesses

Shortly after metric units of length were officially introduced in Australia in the 1970s, each one of 44 students was asked to guess, to the nearest meter, the width of the lecture hall in which they were sitting. Another group of 69 students in the same room was asked to guess the width in feet, to the nearest foot. The measured width of the room was 13.1 meters (43.0 feet). The data, collected by Professor T. Lewis, are given here in Table 2.1, which is taken from Hand et al. (1994). Of primary interest here is whether the guesses made in meters differ from the guesses made in feet, and which set of guesses give the most accurate assessment of the "true" width of the room (accuracy in this context implies guesses which are closer to the measured width of the room).

Table 2.1 Room Width Estimates

```
Guesses in meters
        8     9    10    10    10    10    10    10    11
       11    11    11    12    12    13    13    13    14
       14    14    15    15    15    15    15    15    15
       15    16    16    16    17    17    17    17    18
       18    20    22    25    27    35    38    40
Guesses in feet
       24    25    27    30    30    30    30    30    30
       32    32    33    34    34    34    35    35    36
       36    36    37    37    40    40    40    40    40
       40    40    40    40    41    41    42    42    42
       42    43    43    44    44    44    45    45    45
       45    45    45    46    46    47    48    48    50
       50    50    51    54    54    54    55    55    60
       60    63    70    75    80    94
```

2.2.1 Initial Analysis of Room Width Guesses Using Simple Summary Statistics and Graphics

How should we begin our investigation of the room-width guesses data that are given in Table 2.1? As with most data sets, the initial data analysis steps should involve the calculation of simple summary statistics, such as means and variances, and graphs and diagrams that convey clearly the general characteristics of the data, and perhaps enable any unusual observations or patterns in the data to be detected. Such summary statistics and graphs are very easy to obtain using SAS Enterprise Guide. First, we will show how to read in the data, convert the room widths in meters into feet by multiplying them by 3.28, and then calculate the means and standard deviations of the meter estimates and the feet estimates.

The data are stored in a tab-separated file, **lengths.tab**. To read them in:

1. Select **File≻Import Data**.

2. Select **Local Computer** as the source.

3. Browse to the folder that contains the file, **c:\saseg\data**, select **lengths.tab**, and **Open**. The Import Data window opens.

4. Select **Text Format**, and click the **Delimited** and **Tab** buttons.

5. Select **Column Options**. SAS Enterprise Guide has recognized that the file contains two columns of data, the first character and the second numeric.

6. Uncheck the box **Use column name as label for all columns**.

7. Rename the columns to **units** and **length**. The window should now look like Display 2.1.

8. Under **Results**, click **Browse**, and rename the output file to **SASlengths**.

9. Run the procedure.

Display 2.1 Import Data Task Column Options Pane Room Width Guesses Data

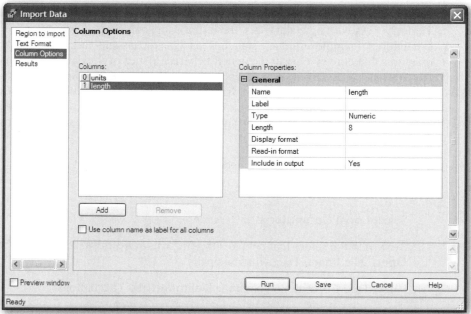

The data are read into a SAS data set, and the cases are visible in the workspace. We can scroll down to check that all cases have been read correctly. Having done this, we can close the view of the data and return to the Process Flow window.

To create a new column with all estimates in feet:

1. Select **Data≻Filter and Query**.

2. In the **Query Builder** window, select **Computed Columns≻New≻Build Expression**. This opens the Advanced Expression Editor.

3. Click the **Functions** tab, select **Conditional** as the function category, select **CASE {short}**, and click **Add to Expression**.

4. Select the first **<whenCondition>** and type **units='m'**. Take care to include a space after what you type so that it does not run into the **THEN** which follows.

5. In the same way, replace the first **<resultExpression>** with **length*3.28**, the second **<whenCondition>** with **units='f'** and the second **<resultExpression>** with **length**, and click **OK**. In each instance, take care to insert a space after what you type.

6. The entire expression should now read **CASE WHEN units='m' THEN length*3.28 WHEN units='f' THEN length END** as shown in Display 2.2. Click **OK**. In the pop-up window, rename **Calculation1** to **feet**, and then **Close**.

Display 2.2 Advanced Expression Editor

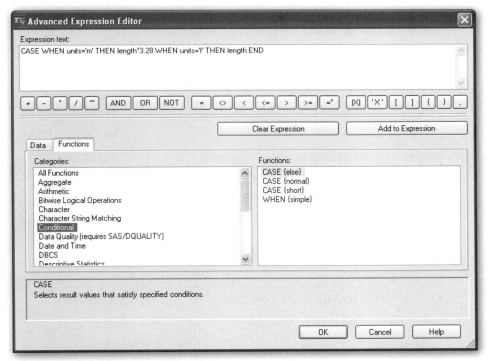

It helps to keep the process flow clear if both the query and the output file are given meaningful names. For example, name the query **Meters2Feet** and the output data set **SASlengths2**. The results appear in the workspace and again we scroll through them to check that they are correct and close the data set.

Deriving Summary Statistics

Summary statistics could be produced with the task of that name (**Describe≻Summary Statistics**) but **Distribution Analysis** is more flexible and produces the graphs that we will use as well as summary statistics.

1. Select **Describe≻Distribution Analysis**.

2. Under **Task Roles**, the **Analysis variable** is **feet**. To compare the summaries for each set of guesses, treat the **units** variable as a **Classification variable**. This generates separate results for each value of **units**.

3. Under **Tables**, select only **Basic measures** for now.

The results are shown in Table 2.2.

Table 2.2 Summary Statistics for Room Width Guesses Data

(a) Guesses made in feet

Basic Statistical Measures			
Location		**Variability**	
Mean	43.69565	Std Deviation	12.49742
Median	42.00000	Variance	156.18542
Mode	40.00000	Range	70.00000
		Interquartile Range	12.00000

(b) Guesses made in meters and then converted to feet

Basic Statistical Measures			
Location		**Variability**	
Mean	52.55455	Std Deviation	23.43444
Median	49.20000	Variance	549.17310
Mode	49.20000	Range	104.96000
		Interquartile Range	19.68000

What do the summary statistics tell us about the two sets of guesses? It appears that the guesses made in feet are closer to the measured room width and less variable than the guesses made in meters suggesting that the guesses made in the more familiar units, feet, are more accurate than those made in the recently introduced units, meters. But often such apparent differences in means and in variation can be traced to the effect of one or two unusual observations that statisticians like to call *outliers*. Such observations can usually be uncovered by some simple graphics, and here we shall construct *box plots* of the two sets of guesses after converting the guesses made in meters to feet.

Constructing Box Plots

A box plot is a graphical display useful for highlighting important distributional features of a continuous measurement. The diagram is based on what is known as the *five-number summary* of a data set, the numbers in question being the minimum, the lower quartile, the median, the upper quartile, and the maximum. The box plot is constructed by first drawing a box with ends at the lower and upper quartiles of the data. Next, a horizontal line (or some other feature) is used to indicate the position of the median within the box, and then lines are drawn from each end of the box to points defined by the upper quartile plus 1.5 times the *interquartile range* (the difference between the upper and lower quartiles) and the lower quartile minus 1.5 times the interquartile range. Any observations outside these limits are represented individually by some means in the finished graphic. Such observations are likely candidates to be labeled *outliers*. The resulting diagram schematically represents the body of the data minus the extreme observations and is particularly useful for comparing the distributional features of a measurement made in different groups.

Distribution analysis also produces box plots, so we can rerun that task to get the plots.

1. In the Process Flow window, reopen the task (double-click its icon or right-click **Open**).

2. In **Plots**, select **Box plot**.

3. Click **Run**.

4. Reply **Yes** to **Would you like to replace the results from the previous run?**

The resulting plots are shown in Figure 2.1; they indicate that both sets of guesses contain a number of possible outliers and also that the guesses made in meters are *skewed* (have a longer tail) and are more variable than the guesses made in feet. We shall return to these findings in the next subsection.

Figure 2.1 Box Plots of Room Width Guesses Made in Feet and in Meters (after Conversion to Feet)

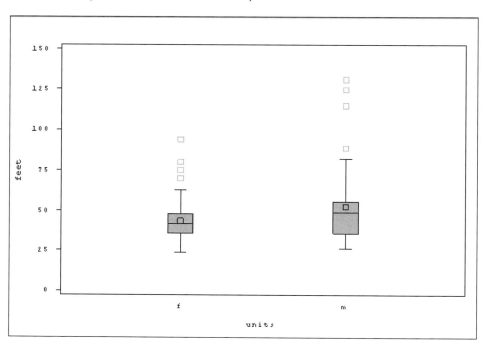

Constructing Histograms and Stem-and-Leaf Plots

The box plot is our favorite graphic for comparing the distributional properties of a measurement made in different groups, but there are other graphics available within **distribution analysis**: *histograms* and *stem-and-leaf plots*. In a histogram, class frequencies are represented by the areas of rectangles centered on the class interval; if class intervals are all equal, then the heights of the rectangles are proportional to the observed frequencies. A stem-and-leaf plot has the shape of the corresponding histogram; but, by also retaining the actual observation values, gives more information. Again, we can rerun the procedure to include these. Stem-and-leaf plots are included in **text based plots**.

The resulting plots are all shown in Figure 2.2; they all show clearly the greater skewness in the guesses made in meters.

Figure 2.2 Histograms and Stem and Leaf Plots for Room Width Guesses Data

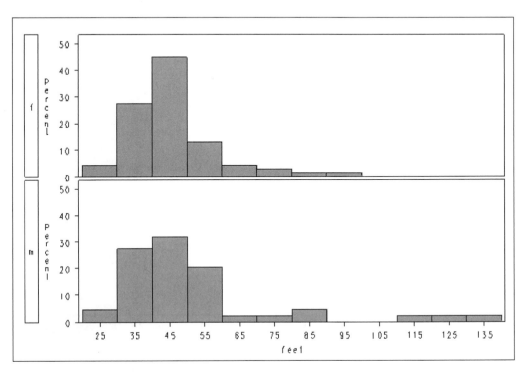

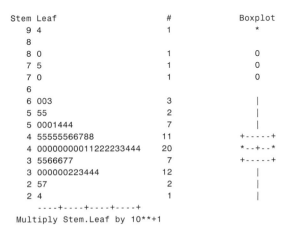

```
     Stem Leaf                          #        Boxplot
       13 1                             1           *
       12 5                             1           *
       12
       11 5                             1           0
       11
       10
       10
        9
        9
        8 9                             1           0
        8 2                             1           |
        7                                           |
        7 2                             1           |
        6 6                             1           |
        6                                           |
        5 666699                        6        +-----+
        5 222                           3        |  +  |
        4 66699999999                  11        *-----*
        4 333                           3        |     |
        3 666699                        6        +-----+
        3 0333333                       7           |
        2 6                             1           |
          ----+----+----+----+
     Multiply Stem.Leaf by 10**+1
```

2.2.2 Guessing the Width of a Room: Is There Any Difference in Guesses Made in Feet and in Meters?

From the summary statistics and graphics produced in the previous section, we already know quite a bit about the how the guesses of room width made in feet differ from the guesses made in meters. The guesses made in feet appear to be concentrated around the measured room width of 43.0 feet; whereas, the guesses made in meters suggest overestimation of the width of the room. In some circumstances, we might simply stop here and try to find an explanation of the apparent difference between the two types of guesses (and many statisticians would be sympathetic to this approach!). But in general the investigation of the data will need to go further and use more formal statistical methods to try to confirm our very strong hunch that guesses of room width made in meters differ from guesses made in feet.

The area of statistics we need to move into is that of *statistical inference*, the process of drawing conclusions about a *population* on the basis of measurements or observations made on a *sample* of observations from the population. This process is central to statistics. More specifically, inference is about testing hypotheses of interest about some population value on the basis of the sample values, and involves what is known as *significance tests*. For the room-width guesses data in Table 2.1, for example, there are three hypotheses we might wish to test:

- In the population of guesses made in meters, the mean is the same as the true room width, namely 13.1 meters. Formally we might write this hypothesis as

$$H_0 : \mu_m = 13.1$$

  where H_0 stands for *null hypothesis.*

- In the population of guesses made in feet, the mean is the same as the true room width namely 43.0 feet; i.e.,

$$H_0 : \mu_f = 43.0$$

- After the conversion of meters into feet, the population means of both types of guess are equal or in formal terms

$$H_0 : \mu_m \mathrm{x} 3.28 = \mu_f$$

It might be imagined that a conclusion about the last of these three hypotheses would be implied from the results found for the first two but, as we shall see later, this is *not* the case.

Applying Student's *t*-Test to the Guesses of Room Width

Testing hypotheses about population means requires what is know as *Student's t-test*. The test is described in detail in Altman (1991), but in essence involves the calculation of a *test statistic* from sample means and standard deviations, the distribution of which is known if the null hypothesis is true and certain assumptions are met. From the known distribution of the test statistic, a *p*-value can be found.

The *p*-value is probably the most ubiquitous statistical index found in the applied sciences literature and is particularly widely used in biomedical and psychological research. So, just what is the *p*-value? Well, the *p*-value is the probability of obtaining the observed data (or data that represent a more extreme departure from the null hypothesis) if the null hypothesis is true, and was first proposed as part of a quasi-formal method of inference by a famous statistician, Ronald Aylmer Fisher, in his influential 1925 book, *Statistical Methods for Research Workers*. For Fisher, the *p*-value represented an attempt to provide a relatively informal measure of evidence against the null hypothesis; the smaller the *p*-value, the greater the evidence that the null hypothesis is incorrect.

But sadly, Fisher's informal approach to interpreting the *p*-value was long ago abandoned in favor of a simple division of results into significant and nonsignificant on the basis of comparing the *p*-value with some largely arbitrary threshold value such as 0.05. The implication of this division is that there can always be a simple "yes" (significant) or "no" (nonsignificant) answer as the fundamental result from a study. This is clearly false.

Used in this way, hypothesis testing is of limited value. In fact, overemphasis on hypothesis testing and the use of *p*-values to dichotomize significant or nonsignificant results has distracted from other more useful approaches to interpreting study results, in particular the use of *confidence intervals*. Such intervals are far more useful alternatives to *p*-values for presenting results in relation to a statistical null hypothesis and give a range of values for a quantity of interest that includes the population value of the quantity with some specified probability. Confidence intervals are described in detail in Altman (1991). In essence, the significance test and associated *p*-value relate to what the population quantity of interest is *not*; the confidence interval gives a plausible range for what the quantity *is*.

So, after this rather lengthy digression, let's apply the relevant Student's *t*-tests to the three hypotheses we are interested in assessing on the room-width data. The first two hypotheses require the application of the single sample *t*-test separately to each set of guesses.

We begin by returning to the Process Flow window with the lengths data by clicking on its tab. To analyze the two sets of guesses separately, we will split the data into two subsets:

1. Click on **SASwaves2** to make it the active data set.

2. Select **Data➤Filter and Query**, click the **Filter Data** tab, and drag **units** across.

3. In the Edit Filter window, type **m** in the value box. Click **OK**. This returns you to the Query Builder window (see Display 2.3). Change the output name to **meters** and click **Run**.

4. Repeat this by typing **f** in the value box and naming the output **feet**.

Display 2.3 Filter Data Selecting Guesses Made in Meters

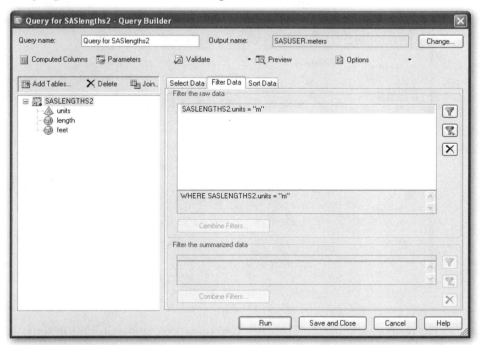

The **t Test** procedure can be used to apply the one sample *t*-test to each set of guesses:

1. Select the **meters** data set.

2. Select **Analyze≻ANOVA≻t Test**.

3. Under **t Test type**, select **One Sample**.

4. Under **Task Roles**, choose **length** as the analysis variable (not **feet** because we want the original units).

5. Under **Analysis**, enter **13.1** for **Specify the test value for the null hypothesis** (Display 2.4).

6. Under **Titles**, amend the title to include **H0=13.1**.

7. Click **Run**.

For the other set of guesses, select the **feet** data set and repeat entering **43** as the test value. Change the title to include **H0=43** and click **Run**.

Display 2.4 Single Sample *t*-Test: Specifying the Value of the Null Hypothesis

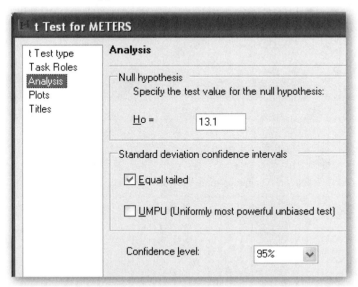

The results are shown in Table 2.3. Let's now look at these results in some detail. Looking first at the two *p*-values, we see that there is no evidence that the guesses made in feet differ in mean from the true width of the room, 43 feet; the 95 % confidence interval here is [40.69,46.70], which includes the true width of the room. But there is considerable evidence that the guesses made in meters do differ from the true value of 13.1 meters; here, the confidence interval is [13.85,18.20], and the students appear to systematically overestimate the width of the room when guessing in meters.

Table 2.3 Results of Single Sample *t*-Tests for Room-Width Guesses Made in Meters and for Guesses Made in Feet

(a) Guesses in meters

Statistics										
Variable	**N**	**Lower CL Mean**	**Mean**	**Upper CL Mean**	**Lower CL Std Dev**	**Std Dev**	**Upper CL Std Dev**	**Std Err**	**Min**	**Max**
Length	44	13.851	16.023	18.195	5.9031	7.1446	9.0525	1.0771	8	40

T-Tests			
Variable	**DF**	**T Value**	**Pr > |t|**
Length	43	2.71	0.0095

(b) Guesses in feet

Statistics										
Variable	**N**	**Lower CL Mean**	**Mean**	**Upper CL Mean**	**Lower CL Std Dev**	**Std Dev**	**Upper CL Std Dev**	**Std Err**	**Min**	**Max**
length	69	40.693	43.696	46.698	10.704	12.497	15.018	1.5045	24	94

T-Tests			
Variable	**DF**	**T Value**	**Pr > |t|**
Length	68	0.46	0.6453

Now, it might be thought that our third hypothesis discussed above, namely that the mean of the guesses made in feet and the mean of the guesses made in meters (after conversion to feet) are the same, can be assessed simply from the results given in Table 2.3. Since the population mean of guesses made in feet apparently does not differ from the true width of the lecture room, but the population mean of guesses made in meters does differ from the true width, can we not simply infer that the population means of the two types of guesses differ from each other? Not necessarily; to assess the equality of means hypothesis correctly, we need to apply an *independent samples t*-test to the data. We again use the *t*-test task.

1. Select the **SASlengths2** data set (click on its icon).

2. Select **Analyze➤ANOVA➤t Test**.

3. Under **t Test type**, select **Two Sample**.

4. Under **Task Roles**, assign **feet** as the **Analysis** variable and **units** as the **Group by** variable (not the **Group analysis by** variable).

5. Click **Run**.

The results of applying this test are shown in Table 2.4. Looking first at the *p*-value when equality of variances is assumed (p=0.0102), we see that there is considerable evidence that the population means of the two types of guesses do indeed differ. The confidence interval for the difference, [−15.57,−2.15], indicates that the guesses made in feet have a mean that is between 16 and 2 feet lower than the guesses made in meters.

Table 2.4 Results of Applying Independent Samples *t*-Test to the Room-Width Guesses Data

Statistics											
Variable	Units	N	Lower CL Mean	Mean	Upper CL Mean	Lower CL Std Dev	Std Dev	Upper CL Std Dev	Std Err	Min
Feet	F	69	40.693	43.696	46.698	10.704	12.497	15.018	1.5045	24
Feet	M	44	45.43	52.555	59.679	19.362	23.434	29.692	3.5329	26.24
Feet	Diff (1-2)		-15.57	-8.859	-2.145	15.524	17.562	20.22	3.3881	

Statistics		
Variable	Units	Maximum
Feet	F	94
Feet	M	131.2
Feet	Diff (1-2)	

T-Tests					
Variable	Method	Variances	DF	t Value	Pr > |t|
Feet	Pooled	Equal	111	-2.61	0.0102
Feet	Satterthwaite	Unequal	58.8	-2.31	0.0246

Equality of Variances					
Variable	Method	Num DF	Den DF	F Value	Pr > F
Feet	Folded F	43	68	3.52	<.0001

2.2.3 Checking the Assumptions Made When Using Student's *t*-Test and Alternatives to the *t*-Test

Having applied *t*-tests to assess each of the hypotheses of interest and having found the corresponding *p*-values and confidence intervals, it might appear that we have finished the analysis of the room-width data. But, as yet, we have not looked at the assumptions that underlie the *t*-tests and have not checked whether these assumptions are likely to be valid for the data. First the assumptions:

- The measurements are assumed to be sampled from a *normal distribution*.
- For the independent samples *t*-test, each population is assumed to have the same variance.
- The measurements made are *independent* of each other.

If any (or all) of these assumptions is invalid, then the *t*-test is not valid strictly speaking. In practice, small departures from the assumptions are unlikely to be of any great importance, but it is still worth trying (informally) to check whether the data meet the assumptions. So let's first consider the normality assumption. One simple way to check for normality is to use what is known as a *probability plot* which, in essence, involves a plot of the ordered observations against theoretical quantiles of the normal distribution (Everitt and Palmer 2006). Such plots should have the form of a straight line; i.e., be *linear* if the sample does arise from a normal distribution. These plots are also available within **Distribution Analysis**, so we can rerun the procedure on the lengths data to obtain them. The resulting plots are shown in Figure 2.3.

Figure 2.3 Probability Plots for the Room Width Guesses Made in Feet and in Meters

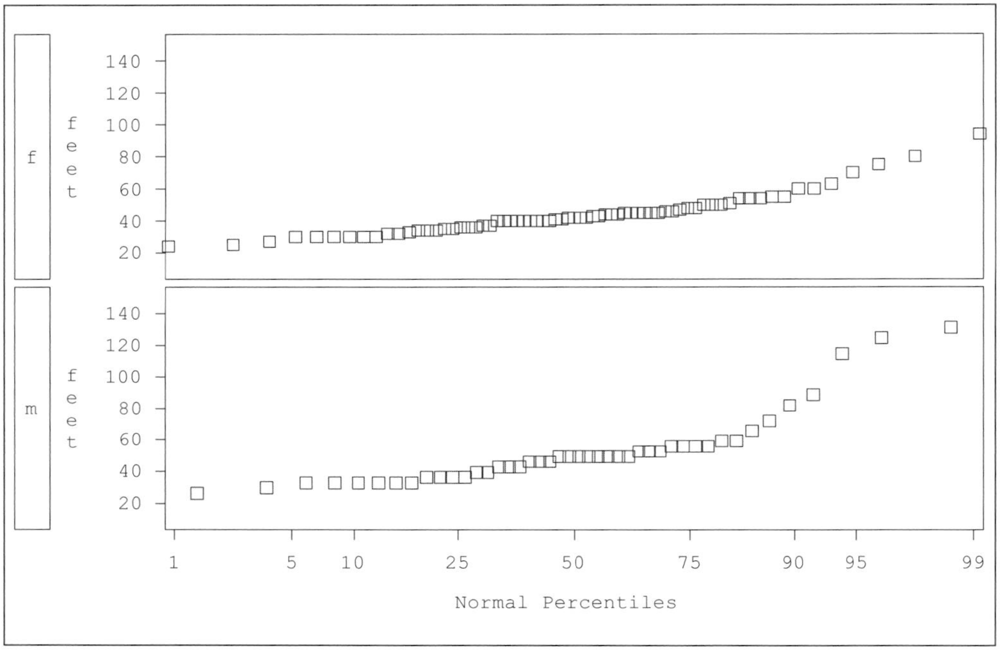

Both plots, but particularly the plot for the guesses in meters, depart from linearity, throwing the normality assumption required for the *t*-test to be valid into some doubt. This possible non-normality—combined with the evidence that two types of guesses have different variances obtained from both the initial examination of the data and the test for equality of variances (Altman 1991) given in Table 2.4—suggests that some caution is needed in interpreting the results from our *t*-tests. Fortunately, the *t*-test is known to be relatively *robust* against departures both from normality and the homogeneity assumption, although it is somewhat difficult to predict how a combination of non-normality, heterogeneity, and outliers will affect the test.

Since the test for equality of variance given in Table 2.4 has an associated *p*-value <0.001, we should perhaps first consider using a modified version of the *t*-test in which the equality of variance assumption is dropped (Altman 1991). The *p*-value of the modified test (*Satterthwaite test*) is also given in Table 2.4 and, although less significant than the usual form of the *t*-test, still shows evidence for a difference in the population means of the two types of room-width guesses.

Here however, given the existence of outliers in the data and their possible non-normality, we might ask whether an alternative test is available that is both insensitive to the effect of outliers and does not assume normality.

Wilcoxon-Mann-Whitney Test

An alternative to Student's *t*-test, which does not depend on the assumption of normality, is the *Wilcoxon-Mann-Whitney test;* this test, since it is based on the ranks of the observations, is also unlikely to be affected greatly by outliers. The Wilcoxon-Mann-Whitney test, which is described in detail in Altman (1991), assesses whether the distribution of the measurements in the two groups are the same. We can apply the test here as follows:

1. Select **Analyze➢ANOVA➢Nonparametric One-Way Anova**.

2. Under **Task Roles** assign **feet** the role of **Dependent** variable, and assign **units** that of **Independent** variable.

3. Under **Analysis**, select only **Wilcoxon**, and uncheck the other options.

4. Click **Run**.

The *p*-value for the test is 0.028 confirming the difference in location between the guesses in feet and the guesses in meters.

2.3 Example: Wave Power and Mooring Methods

In a design study for a device to generate electricity from wave power at sea, experiments were carried out on scale models in a wave tank to establish how the choice of mooring method for the system affected the bending stress produced in part of the device. The wave tank could simulate a wide range of sea states (rough, calm, moderate, etc.) and the model system was subjected to the same sample of sea states with each of two mooring methods, one of which was considerably cheaper than the other. The resulting data giving root mean square bending moment in Newton meters are shown in Table 2.5. These data are taken from Hand et al. (1994). The question of interest is whether bending stress differs for the two mooring methods.

Table 2.5 Wave Energy Device Mooring Data

Sea State	Method I	Method II
1	2.23	1.82
2	2.55	2.42
3	7.99	8.26
4	4.09	3.46
5	9.62	9.77
6	1.59	1.40
7	8.98	8.88
8	0.82	0.87
9	10.83	11.20
10	1.54	1.33
11	10.75	10.32
12	5.79	5.87
13	5.91	6.44
14	5.79	5.87
15	5.50	5.30
16	9.96	9.82
17	1.92	1.69
18	7.38	7.41

2.3.1 Initial Analysis of Wave Energy Data Using Box Plots

For the wave energy data in Table 2.5, we will construct box plots of the bending stresses for each mooring method and, for reasons which will become apparent in the next subsection, it is also useful to have a look at the box plot of the differences between the pairs of observations made for the same sea state.

To keep the analyses of the two examples separate, we open a new Process Flow window for the waves data.

1. Select **File≻New≻Process Flow**.

2. Rename this new process flow **Waves** (right-click on the tab and select **Rename**). We could also rename the other process flow **Lengths** at this point.

The data are stored in a tab-separated file, **waves.tab**. To read them in:

1. Select **File≻Import Data**.

2. Select **Local Computer** as the source.

3. Then browse to the folder that contains the file, **c:\saseg\data**, select **waves.tab**, and click **Open**. The Import Data window opens.

4. Under **Text Format**, click the **Delimited** and **Tab** buttons.

5. Under **Column Options**, SAS Enterprise Guide has recognized that the file contains three columns of numeric data. We rename these to **pairno**, **rsmb1**, and **rsmb2**.

6. Under **Results**, change the name of the output data set to **SASwaves**.

7. Click **Run**.

The data are read into a SAS data set and are shown in the workspace.

To create a new variable with the differences:

1. Select **Data≻Filter and Query**.

2. In the **Query Builder** window, select **Computed Columns≻New≻Build Expression**.

3. In the **Advanced Expression Editor** in the **Expression text** window, type **rsmb1 − rsmb2**, and click **OK**.

4. In the **Computed** window, rename **Calculation1** to **difference**, and then **Close**.

5. Name the query **calc_difference** and the output data set **SASwaves2**.

6. **Run** the query.

Distribution Analysis (**Describe≻Distribution Analysis**) is used to produce box plots for **rsmb1**, **rsmb2** (see Figure 2.4), and **difference** (see Figure 2.5). All three are assigned the roles of **Analysis variables**.

Figure 2.4 Box Plots of Root Mean Square Bending Moment (Newton Meters) for Mooring Methods I and II

(a) Method I

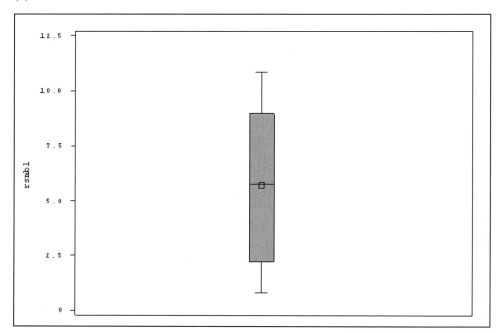

(b) Method II

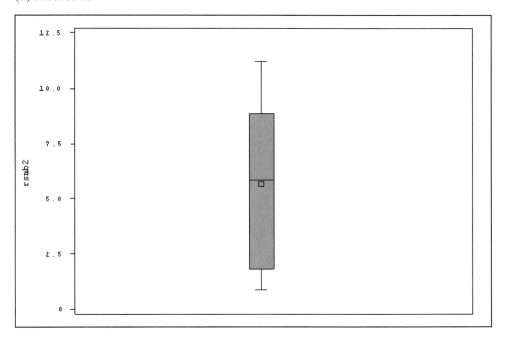

Figure 2.5 Box Plot of Differences of Root Mean Square Bending Moment for the Two Mooring Methods

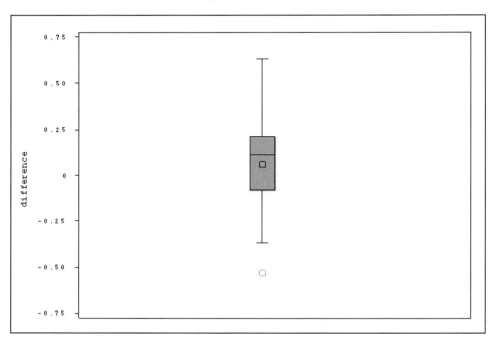

The box plot of differences in Figure 2.5 suggests that there may be one outlying observation that we may wish to check, and a small degree of skewness—although with only 18 observations, drawing any conclusions about the distributional properties of the data is difficult.

2.3.2 Wave Power and Mooring Methods: Do Two Mooring Methods Differ in Bending Stress?

Now, we can move on to consider more formally the questions of interest about the wave energy data. Superficially, these data look to be of a very similar format to the room-width guesses data, but closer consideration shows that there is a fundamental difference in that the observations are paired; i.e., the bending stress for each of the two mooring methods is, in each case, based on the same sea state. Consequently, these observations are more likely to be correlated rather than independent. To test whether there is a difference in the mean bending stress of the two methods of mooring, we use what is called a *paired t*-test (Altman 1991); essentially, this test is the same as the single sample *t*-test used previously for the room width data but here the null hypothesis is that the population mean of the differences of the paired observations is zero. To apply the test:

1. Select the **SASwaves2** data set.

2. Select **Analyze≻ANOVA≻t Test**.

3. Under **t Test type**, choose **Paired**.

4. Under **Task Roles**, assign **rsmb1** and **rmsb2** as the **Paired variables** (Display 2.5).

5. **Run** the procedure.

Display 2.5 Task Roles Pane for Matched Pairs *t*-Test

The results are shown in Table 2.6. The *p*-value for the test is 0.38, and the 95% confidence interval for the mean of the differences is [–0.083, 0.206]. There is no evidence of any difference in the mean bending stress of the two mooring methods.

Table 2.6 Paired *t*-Test for Wave Energy Mooring Data

Statistics										
Difference	N	Lower CL Mean	Mean	Upper CL Mean	Lower CL Std Dev	Std Dev	Upper CL Std Dev	Std Err	Min
rmsb1 - rmsb2	18	-0.083	0.0617	0.2059	0.2177	0.2901	0.4349	0.0684	0.53

Statistics	
Difference	Maximum
rmsb1 - rmsb2	0.63

T-Tests			
Difference	DF	T Value	Pr > |t|
rmsb1 - rmsb2	17	0.90	0.3797

2.3.3 Checking the Assumptions of the Paired *t*-Tests

For the paired *t*-test to be valid, the differences between the paired observations need to be normally distributed. We could use a probability plot to assess the required normality of the differences but, with only 18 observations for the wave data, the plot would not be very useful. Since we cannot satisfactorily assess the normality assumption for these data, we might wish to consider a nonparametric alternative; this would be the *Wilcoxon signed rank test*.

Wilcoxon Signed Rank Test

The non-parametric analogue of the paired *t*-test is Wilcoxon's signed rank test (Altman 1991). As with the Wilcoxon-Mann-Whitney test described above, the signed rank test uses only the ranks of the observations and does not assume normality for the observations; the test can be applied within **Distribution Analysis**.

1. Select **Analyze▷Distribution Analysis**.

2. Under **Task Roles**, assign **difference** as the **Analysis** variable.

3. Under **Tables**, select **Tests for location**.

This is also an alternative way of applying a matched pairs *t*-test, as can be seen from the results in Table 2.7.

Table 2.7 Wilcoxon Signed Rank Test for Wave Energy Mooring Data

Tests for Location: Mu0=0			
Test		Statistic	p Value
Student's t	T	0.90193	Pr > |t| 0.3797
		4	
Sign	M	1	Pr >= |M| 0.8145
Signed Rank	S	23.5	Pr >= |S| 0.3194

The test gives a *p*-value of 0.319 confirming the result from the paired *t*-test.

2.4 Exercises

Exercise 2.1 The **babies** data set gives the recorded birthweights of 50 infants who displayed severe idiopathic respiratory distress syndrome (SIRDS). SIRDS is a serious condition that can result in death and did so in the case of 27 of these children. One of the questions of interest about these data is whether the babies who died differed in birthweight from the babies who survived. Use some suitable graphical techniques to carry out an initial analysis of these data and then find a 95% confidence interval for the difference in mean birthweight for SIRDS babies who die and SIRDS babies who live.

```
Birthweights (kg)

Survived
1.130   1.575   1.680   1.760   1.930   2.015   2.090   2.600
2.700   2.950   3.160   3.400   3.640   2.830   1.410   1.715
1.720   2.040   2.200   2.400   2.550   2.570   3.005

Died
1.050   1.175   1.230   1.310   1.500   1.600   1.720   1.750
1.770   2.275   2.500   1.030   1.100   1.185   1.225   1.262
1.295   1.300   1.550   1.820   1.890   1.940   2.200   2.270
2.440   2.560   2.370
```

Exercise 2.2 The data in the **choles** data set were collected by the Western Collaborative Group Study carried out in California in 1960–1961. In this study, 3,154 middle-aged men were used to investigate the possible relationship between behavior pattern and the risk of coronary heart disease. The data set contains data from the 38 heaviest men in the study (all weighing at least 225 pounds). Cholesterol measurements (mg per 100ml) and behavior type were recorded; type A behavior is characterized by urgency, aggression, and ambition; type B behavior is relaxed, non-competitive, and less hurried. The question of interest is whether, in heavy middle-aged men, cholesterol level is related to behavior type. Investigate the question of interest in any way you feel is appropriate, paying particular attention to assumptions and to any observations that might possibly distort conclusions.

Type A:

233 291 312 250 246 197 268 224 329 239 254 276 234 181 248 252 202 218 325

Type B:

420 185 263 246 224 212 188 250 148 169 226 175 242 153 183 137 202 194 213

Exercise 2.3 The data in **diet** come from a study of the Stillman diet, a diet that consists primarily of protein and animal fats, and restricts carbohydrate intake. In **diet**, triglyceride values (mg/100ml) are given for 16 participants both before beginning the diet and at the end of a period of time following the diet. Here, interest is on whether there has been a change in triglyceride level that might be attributed to the diet. Carry out an appropriate hypothesis test to investigate whether there has been a change in triglyceride level using any graphics that you think might be helpful in interpreting the test.

Subject	Baseline	Final
1	159	194
2	93	122
3	130	158
4	174	154
5	148	93
6	148	90
7	85	101
8	180	99
9	92	183
10	89	82
11	204	100
12	182	104
13	110	72
14	88	108
15	134	110
16	84	81

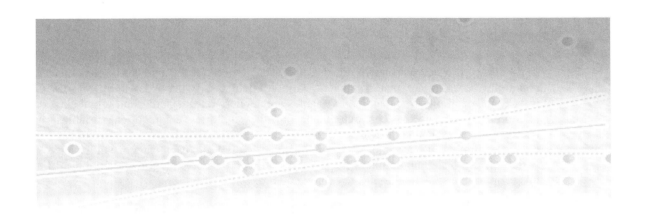

Chapter **3**

Dealing with Categorical Data

3.1 Introduction

In this chapter, we discuss how to deal with various aspects of the analysis of data containing *categorical variables*; that is, variables that classify the observations in some way. Some examples of categorical variables are gender, marital status, and social class. Numbers might be used as convenient labels for the categories of categorical variables but have no numerical significance. When using categorical variables, we can simply count the number of our sample—or how many—that fall into each category of a variable, or into a combination of the categories of two or more categorical variables. In this chapter, the statistical topics to be covered are:

- Graphical summary of one-way tables, bar charts, and pie charts
- Testing for association of two categorical variables—Chi-square tests for independence
- Testing for association of two categorical variables when some observed counts are small—Fisher's exact test
- Testing for equal probability of an event in matched pairs data—McNemar's test

3.2 Example: Horse Race Winners

The data shown in Table 3.1 show the starting stalls of the winners in 144 horse races held in the U.S. All 144 races took place on a circular track and all races relate to races with eight horses each. Starting stall 1 is closest to the rail on the inside of the track. Interest here lies in assessing how the chances of a horse winning a race are affected by its position in the starting lineup.

Table 3.1 Horse Racing Data after Classification

Starting stall	1	2	3	4	5	6	7	8
Number of winners	29	19	18	25	17	10	15	11

3.2.1 Looking at Horse Race Winners Using Some Simple Graphics: Bar Charts and Pie Charts

The horse racing data are in a SAS data set, **racestalls**, which contains a single variable giving the stall number for each of the 144 winners. To add the data set to the project:

1. Select **File≻Open≻Data≻Local Computer**.

2. Browse to the folder containing the SAS data sets, **c:\saseg\sasdata**, select **racestalls.sas7bdat**, and **Open**.

We can now reproduce Table 3.1 showing the number of winners from each of the eight starting stalls and the corresponding percentages using:

1. Select **Describe≻One-Way Frequencies**.

2. Under **Task Roles**, select the only variable, **stall**, as the **Analysis** variable.

3. Click **Run**.

The result is shown in Table 3.2. We see that the percentage of winning horses from each stall differs considerably suggesting that *stall* does play a part in determining which horse will win.

Table 3.2 Horse Racing Data

Stall	Frequency	Percent	Cumulative Frequency	Cumulative Percent
1	29	20.14	29	20.14
2	19	13.19	48	33.33
3	18	12.50	66	45.83
4	25	17.36	91	63.19
5	17	11.81	108	75.00
6	10	6.94	118	81.94
7	15	10.42	133	92.36
8	11	7.64	144	100.00

The counts (or percentages) in Table 3.2 can be represented graphically by a *bar chart* (**Graph≻Bar Chart)** or a *pie chart* (**Graph≻Pie Chart**). Bar charts are also available via **Describe≻One-Way Frequencies**. To produce a bar chart in this way:

1. Reopen the **One-Way Frequencies** task in the **Process Flow** window (double-click on the icon, or right-click **Open**).

2. Under **Plots**, select **horizontal bar charts**.

3. Click **Run**.

4. Answer **Yes** to **Would you like to replace the results from the previous run?**

For the pie chart, select **Graph≻Pie Chart**, assign **stall** the role of **Column to chart**, and click **Run**.

The resulting diagrams are shown in Figure 3.1. It should be pointed out that despite their widespread popularity, both the general and scientific use of pie charts have been severely criticized (Tufte 1983 and Cleveland 1994). Both diagrams simply mirror what we previously gleaned from the percentages in Table 3.2, namely that there does appear to be a difference in the number of winners from each stall.

Figure 3.1 Bar Chart and Pie Chart for Horse Racing Data

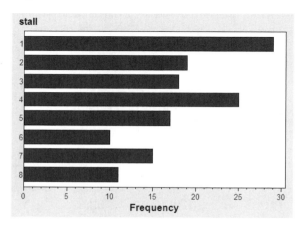

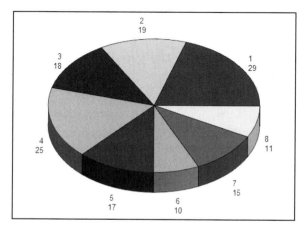

The bar chart often becomes more useful if the bars are arranged in ascending or descending order of frequency. If the format of the graphs produced by SAS Enterprise

Guide is **ActiveX** (to check or select this format **Tools**➢ **Options**➢ **Results**➢ **Graph**➢ **Graph Format**), this can be done interactively.

1. In the output from **One-Way Frequencies**, right-click on the bar chart and select **Data Options**.

2. In the **Data Options** window, under **Vertical axis**, select **Descending Statistic** as the value for **Sort by** (Display 3.1).

3. Click **OK**.

Display 3.1 Using the Data Options Window to Reorder the Bars of a Horizontal Bar Chart

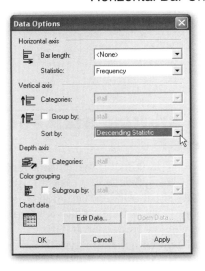

The resulting plot is shown in Figure 3.2. We can now see clearly that starting stalls 1 to 4 produce many more winners than stalls 5 to 8, and starting stall 1 produces the highest number of winners of all eight starting stalls.

Figure 3.2 Ordered Bar Chart for Horse Racing Data

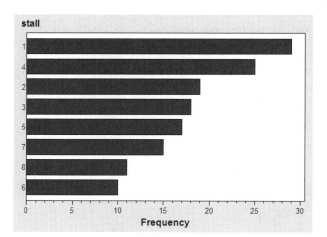

3.2.2 Horse Race Winners: Does Starting Stall Position Predict Horse Race Winners?

What would we expect the counts in Table 3.1 to look like if the starting stall does *not* affect the chances of a horse winning a race? Clearly, we would expect to see the number of winners from each stall to be approximately equal (random variation will stop them being exactly equal). So here our null hypothesis about the population of horse race winners is that there are an equal number of winners from each stall. In our sample of 144 winners, the counts do not appear to be consistent with the null hypothesis but how can we assess the evidence against the null hypothesis formally?

We begin by calculating the counts of winners in each stall we might expect when we observe the results of 144 races, if the null hypothesis is true. We then compare these *expected values* with the observed values using what is known as the *chi-square test statistic*. The expected values for each stall under the null hypothesis are simply 144/8=18, and the chi-square statistic is then calculated as the sum of the square of each difference between the observed and expected value divided by the expected value. So in detail, the required chi-square test statistic is calculated thus:

$$\frac{(29\text{-}18)^2}{18}+\frac{(19-18)^2}{18}+\frac{(18-18)^2}{18}+\frac{(25-18)^2}{18}+\frac{(17-18)^2}{18}+\frac{(10-18)^2}{18}+\frac{(15-18)^2}{18}+\frac{(11-18)^2}{18}$$

If the null hypothesis is true, the chi-square test statistic has a *chi-squared distribution* with seven *degrees of freedom*. Altman (1991) includes full details of the chi-square test.

To apply the test:

1. Reopen the **One-Way Frequencies** task (in the **Process Flow** window; double-click on its icon or right-click **Open**).

2. Under **Statistics,** check **Asymptotic test** in the **Chi-square goodness of fit** box (Display 3.2).

3. Click **Run**.

Display 3.2 Selecting the Chi-Square Test for the Race Stalls Data

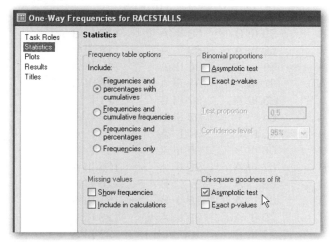

The results are shown in Table 3.3. The chi-square statistic takes the value 16.3 with an associated *p*-value of 0.02. Consequently, there is evidence that *starting stall* is a factor in determining the winning horse, as previously suggested by examination of the frequencies and the corresponding bar charts.

Table 3.3 Chi-Square Test for the Horse Racing Data

Chi-Square Test for Equal Proportions	
Chi-Square	16.3333
DF	7
Pr > ChiSq	0.0222

3.3 Example: Brain Tumors

In an investigation of brain tumors, the type and site of the tumor for 141 individuals were noted. The three possible types were A: benign tumors, B: malignant tumors, and C: other cerebral tumors. The sites of concerned were I: frontal lobes, II: temporal lobes, and III: other cerebral areas. The data are shown in Table 3.4. Do these data give any evidence that some types of tumors occur more frequently at particular sites; i.e., that there is an association between the categorical **type** and **site** variables?

Table 3.4 Data on Type and Site of Brain Tumors

1 III A	30 III B	59 III B	88 II A	117 II B
2 III C	31 II C	60 III A	89 I A	118 III B
3 II A	32 III A	61 II A	90 III A	119 II A
4 I A	33 II A	62 III A	91 III A	120 III C
5 III A	34 II A	63 III A	92 III B	121 I C
6 III C	35 I A	64 I A	93 III C	122 I A
7 I A	36 III B	65 II C	94 I A	123 I C
8 I A	37 II B	66 III B	95 III A	124 I A
9 III A	38 II B	67 III A	96 II A	125 III A
10 III A	39 I B	68 I A	97 I B	126 III A
11 III A	40 III B	69 I A	98 II B	127 III B
12 I A	41 I C	70 II A	99 II A	128 III B
13 III A	42 I A	71 III B	100 III B	129 III A
14 III B	43 I B	72 I C	101 III B	130 III B
15 III A	44 II A	73 II A	102 III C	131 III B
16 III B	45 III B	74 III C	103 I A	132 III A
17 II A	46 II A	75 I A	104 III C	133 III C
18 III A	47 II A	76 II A	105 III A	134 III C
19 I B	48 III A	77 III A	106 III A	135 III B
20 III C	49 I B	78 III C	107 II A	136 III A
21 I A	50 III C	79 III A	108 I C	137 I A
22 III A	51 III B	80 I A	109 III A	138 I B
23 III A	52 III C	81 II A	110 III C	139 III B
24 III A	53 III A	82 I A	111 II A	140 II A
25 III A	54 I A	83 III B	112 III B	141 I A
26 III B	55 III C	84 II C	113 III C	
27 III B	56 III C	85 I C	114 II A	
28 II A	57 III A	86 I A	115 I B	
29 I B	58 III A	87 I A	116 I B	

3.3.1 Tabulating the Brain Tumor Data into a Contingency Table

For the data about brain tumors in Table 3.4, we can cross-classify the observations to give what is know as a 3 x 3 *contingency table* showing the counts in all nine possible combinations of the type and site of tumors categories. The original data are in a SAS data set, **tumors**. Add this to the project, as above, and:

1. Select **Describe≻Table Analysis**.

2. Under **Task Roles**, the two variables **site** and **type** are assigned as **Table variables**.

3. Under **Tables**, drag **type** across to the preview pane and then **site**.

4. The **Tables to be generated** pane should now contain **site by type** as its first line (Display 3.3).

5. Click **Run**.

Display 3.3 Tables Analysis Table Preview Pane

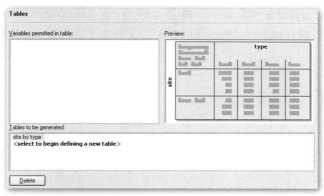

The resulting contingency table is shown in Table 3.5.

Table 3.5 Brain Tumor Data after Cross-Classification

Table of site by type				
Site	**Type**			
Frequency Col Pct	**A**	**B**	**C**	**Total**
I	23	9	6	38
	29.49	24.32	23.08	
II	21	4	3	28
	26.92	10.81	11.54	
III	34	24	17	75
	43.59	64.86	65.38	
Total	78	37	26	141

3.3.2 Do Different Types of Brain Tumors Occur More Frequently at Particular Sites? The Chi-Square Test

We are now interested in assessing the null hypothesis that site of tumor and type of tumor are *independent*. Independence implies that the probabilities of the tumor types are the same at all sites. More explicitly, independence implies that the probability of a patient having a tumor of a particular type at a particular site is simply the product of the probability of this type of tumor multiplied by the probability of a tumor at this site.

We can estimate both the probability of type of tumor and the probability of a tumor at a particular site by simply dividing the appropriate *marginal total* by the number of observations. For example, the estimate of the probability of being a type A tumor is 78/141=0.553, and the estimate of a tumor being at site I is 38/141=0.270. So, if the null hypothesis of independence is true, then the estimate of the probability of a patient having an A type tumor at site I is 0.553 x 0.270=0.149. So, under the assumption of independence, the expected count in the type A, site I cell of the contingency table is 141 x 0.149=21.0. In the same way, we can calculate the expected values for all the other cells in the table and these can then be compared with the observed values by means of the chi-square statistic. For a contingency table with r rows and c columns, the chi-square test of independence has $(r-1)(c-1)$ degrees of freedom where r is the number of rows of the table and c is the number of columns. In the tumor example, both r and c have the value 3 so the chi-square statistic will have four degrees of freedom. Everitt (1992) provides full details of the chi-square test of independence in contingency tables.

The chi-square test is one of the many tests available within **Table Analysis**. To apply it:

1. Open the **Table Analysis** task (double-click or right-click **Open**).

2. Under **Table Statistics≻Association**, check **Chi-square tests** under **Tests of association**.

3. Click **Run**.

4. Replace the results from the previous run.

The result is shown in Table 3.6. Here, the chi-square test statistic takes the value 7.8 and has an associated *p*-value of 0.098; there is no strong evidence against the hypothesis that type and site of tumor are independent. The result implies that the observed values in Table 3.5 do not differ greatly from the corresponding values to be expected if tumor site and type of tumor are independent. Everitt (1992) describes the other terms in Table 3.6.

Table 3.6 Chi-Square Test of Independence for Brain Tumor Data

Chi-Square Tests Statistic	DF	Value	Prob
Chi-Square	4	7.8441	0.0975
Likelihood Ratio Chi-Square	4	8.0958	0.0881
Mantel-Haenszel Chi-Square	1	2.9753	0.0845
Phi Coefficient		0.2359	
Contingency Coefficient		0.2296	
Cramer's V		0.1668	

Sample Size = 141

3.4 Example: Suicides and Baiting Behavior

Mann (1981) reports a study carried out to investigate the causes of jeering or baiting behavior by a crowd when a person is threatening to commit suicide by jumping from a high building. A hypothesis is that baiting is more likely to occur in warm weather. Mann classified 21 accounts of threatened suicide by two factors: the time of the year and whether or not baiting occurred. The classified data are given in Table 3.7 and the question is: Do the data give any evidence to support the "warm weather" hypothesis?

(The data come from the northern hemisphere, so the months June to September are the warm months.)

Table 3.7 Crowd Behavior at Threatened Suicides

	Baiting	Nonbaiting
June–September	8	4
October–May	2	7

3.4.1 How Is Baiting Behavior at Suicides Affected by Season? Fisher's Exact Test

The chi-square test carried out in the previous section for the brain tumor data above depends on knowing that the test statistic has a chi-squared distribution if the null hypothesis of independence is true; this allows *p*-values to be found. But what was not mentioned previously is that the chi-squared distribution is appropriate only under the assumption that the expected values are not "too small." Such a term is almost as vague as asking how long is a piece of string, and has been interpreted in a number of ways. Most commonly, it has been taken as meaning that the chi-squared distribution is appropriate only if all the expected values are five or more. Such a "rule" is widely quoted but appears to have little mathematical or empirical justification over, say, a one-or-more rule.

Nevertheless, for contingency tables based on small sample sizes, the usual form of the chi-square test for independence may not be strictly valid although it is often difficult to predict *a priori* whether a given data set may cause problems. But there may be occasions where it is advisable to consider another approach that is available and that is a test that does not depend of the chi-squared distribution at all. Such *exact* tests of independence for a general r x c contingency table are computationally intensive and, until relatively recently, the computational difficulties have severely limited their application. But within the last ten years, the advent of fast algorithms and the availability of inexpensive computing power have considerably extended the bounds where exact test are feasible. Details of the algorithms for applying exact tests are outside the level of this text, and interested readers are referred to Mehta and Patel (1986a, 1986b) for a full exposition. But for a table in which both r and $c = 2$, there is an exact test which has been in use for decades, namely *Fisher's exact test,* a test that is described in Everitt (1992). Fisher's test is produced by default as part of **Chi-square tests** for a 2 x 2 contingency table. (For larger tables, it is available as an option.)

The data on baiting behavior at suicides provides us with an example of how to use SAS Enterprise Guide to apply Fisher's exact test for a 2 x 2 table and will also serve to illustrate how to analyze data that is in the form of a table rather than individual observations. We begin by creating a new data set to enter the data into:

1. Select **File≻New≻Data**.

2. When prompted, type the name **baiting**.

A data table opens, and we enter the data with one row per cell and a column each for the number in the cell, whether or not there was baiting and whether the season was warm or cool. The columns can be renamed as **baiting**, **season**, and **count**, by right-clicking on the head of the column selecting properties and typing in a new name. The result should look like Display 3.4.

Display 3.4 Baiting Data Entered Directly Into SAS Enterprise Guide

	baiting	season	count
1	yes	warm	8
2	no	warm	4
3	yes	cool	2
4	no	cool	7
5			

Data entered in this way are stored in a temporary data set. When leaving SAS Enterprise Guide, there is the option to discard them or move them to a location where they can be retained. We have saved them to **c:\saseg\sasdata**.

To apply the chi-square test and Fisher's exact test:

1. Select **Describe≻Table Analysis**.

2. Answer **Yes** to protect the data.

3. Under **Task Roles**, **baiting** and **season** are assigned as **Table variables** and **count** as a **Frequency count**.

4. Under **Tables**, drag **baiting** across to the preview pane and then **season**. The **Tables to be generated** pane should now contain **season by baiting** as its first line.

5. Under **Table statistics≻Association**, check **Chi-square tests**.

6. Click **Run**.

The crosstabulation is not reproduced exactly as entered; the categories of **season** and **baiting** are in alphabetical order. It is easier to check that the data have been correctly entered when the table is reproduced as entered. To do this, we could rerun the task and, under **Table Statistics≻Computation Options**, select **Order values by: Data set order.**

The result is shown in Table 3.8. The *p*-value from Fisher's exact test is 0.0805. There is no strong evidence of crowd behavior being associated with the time of year of the threatened suicide, but it has to be remembered that the sample size is low and the test lacks power. (Carrying out the usual chi-square test on these data gives a *p*-value of 0.0436, a considerable difference from the value for Fisher's exact test, and suggesting there *is* evidence of an association between crowd behavior and time of year of threatened suicide.)

Table 3.8 Analysis of Baiting and Suicide Data

Table of season by baiting			
Season	**baiting**		
Frequency **Col Pct**	**no**	**yes**	**Total**
cool	7 63.64	2 20.00	9
warm	4 36.36	8 80.00	12
Total	11	10	21

Chi-Square Tests Statistic	DF	Value	Prob
Chi-Square	1	4.0727	0.0436
Likelihood Ratio Chi-Square	1	4.2535	0.0392
Continuity Adj. Chi-Square	1	2.4858	0.1149
Mantel-Haenszel Chi-Square	1	3.8788	0.0489
Phi Coefficient		0.4404	
Contingency Coefficient		0.4030	
Cramer's V		0.4404	
WARNING: 50% of the cells have expected counts less than 5. Chi-Square may not be a valid test.			

Fisher's Exact Test	
Cell (1,1) Frequency (F)	7
Left-sided Pr <= F	0.9942
Right-sided Pr >= F	0.0563
Table Probability (P)	0.0505
Two-sided Pr <= P	0.0805

3.5 Example: Juvenile Felons

The data in Table 3.9 (Agresti 1996) arise from a sample of juveniles convicted of felony in Florida in 1987. *Matched pairs* of offenders were formed using criteria such as age and number of previous offences. For each pair, one subject was handled in the juvenile court, and the other was transferred to the adult court. Whether or not the juvenile was re-arrested by the end of 1988 was then noted. Here, the question of interest in whether the population proportions re-arrested are identical for the adult and juvenile courts?

Table 3.9 Re-Arrests of Juvenile Felons by Type of Court in Which They Were Tried

	Juvenile court	
Adult court	**Re-arrest**	**No re-arrest**
Re-arrest	158	515
No re-arrest	290	1134

3.5.1 Juvenile Felons: Where Should They Be Tried? McNemar's Test

The chi-square test on categorical data described previously assumes that the observations are independent of one another. But the data on juvenile felons in Table 3.4 arise from matched pairs and so they are *not* independent. The counts in the corresponding 2 x 2 table of the data refer to the pairs, so for example, in 158 of the pairs of offenders, *both* members of the pair were re-arrested. To test whether the re-arrest rate differs between the adult and juvenile courts, we need to apply what is known as

McNemar's test. The test is described in Everitt (1992); to apply it to the juvenile offenders data, we can enter the data directly as with the previous example.

1. Select **File≻New≻Data.**

2. When prompted, supply the name **arrests**. A data table opens and we enter the data one row per cell and a column each for the number in the cell, the adult court outcome, and the juvenile court outcome.

3. Rename the columns as **adult**, **juvenile**, and **count** by right-clicking on the head of the column, selecting **Column Properties**, and typing in a new name. In the example below (Display 3.5), we have entered **re** for re-arrests and **no** for no re-arrests.

Display 3.5 Re-Arrest Data for Juvenile Felons Entered Directly into SAS Enterprise Guide

	adult	juvenile	count
1	re	re	158
2	re	no	515
3	no	re	290
4	no	no	1134

The data have also been stored in **c:\saseg\sasdata**.

To apply McNemar's test:

1. Select **Describe≻Table Analysis**.

2. Answer **Yes** to protect the data.

3. Under **Task Roles**, **adult** and **juvenile** are assigned as **Table variables** and **count** as a **Frequency count**.

4. Under **Tables**, drag **adult** across to the **Preview** pane and then **juvenile**. The **Tables to be generated** pane should now contain **juvenile by adult** as its first line.

5. McNemar's test is located under **Table Statistics≻Agreement**; check **Measures**.

6. To reproduce the table as entered, under **Table Statistics≻Computation Options**, set the option **Order values by:** to **Data set order**.

7. Click **Run**.

The result is shown in Table 3.10. The test statistics takes the value 62.89 with an extremely small associated *p*-value. There is very strong evidence that *type of court* and the *probability of re-arrest* are related. It appears that trial at a juvenile court is less likely to result in re-arrest.

Table 3.10 McNemar's Test for Juvenile Crime Data

<table>
<tr><td colspan="4">Table of adult by juvenile</td></tr>
<tr><td>adult</td><td colspan="3">juvenile</td></tr>
<tr><td>Frequency
Col Pct</td><td>re</td><td>no</td><td>Total</td></tr>
<tr><td>re</td><td>158
35.27</td><td>515
31.23</td><td>673</td></tr>
<tr><td>no</td><td>290
64.73</td><td>1134
68.77</td><td>1424</td></tr>
<tr><td>Total</td><td>448</td><td>1649</td><td>2097</td></tr>
</table>

McNemar's Test	
Statistic (S)	62.8882
DF	1
Pr > S	<.0001

3.6 Exercises

Exercise 3.1 The **crash** data set lists fictitious counts of fatal air crashes in Australia by quarter over a twenty-year period. Assess the hypothesis that the accident rates are uniform across these four quarters:

Jan	April	July	October
12	**8**	**7**	**8**

Exercise 3.2 One hundred American citizens were surveyed and asked to identify which of five items were most fearful to them. The results are given in the **fear** data set. Test whether sex and greatest fear are independent of each other.

	Public speaking	Heights	Insects	Financial Problems	Sickness/Death
Male	12	5	4	17	10
Female	11	15	10	4	12

Exercise 3.3 In a broad general sense, psychiatric patients can be classified as psychotics or neurotics. A psychiatrist, whilst studying the symptoms of a random sample of 20 patients from each type, found that whereas six patients in the neurotic group had suicidal feelings, only two in the psychotic group suffered in this way. Is there any evidence of an association between type of patient and suicidal feelings?

	Psychotics	Neurotics
Suicidal feelings	2	6
No suicidal feelings	18	14

The data are in the **suicidal** data set.

Exercise 3.4 The data in the **cancer** data set arise from an investigation of the frequency of exposure to oral conjugated estrogens among 183 cases of endometrial cancer. Each case was matched on age, race, date of admission, and hospital of admission to a suitable control not suffering from cancer. Is there any evidence that use of oral conjugated estrogens is associated with endometrial cancer?

		Controls	
		Used	Not Used
Cases	Used	12	43
	Not used	7	121

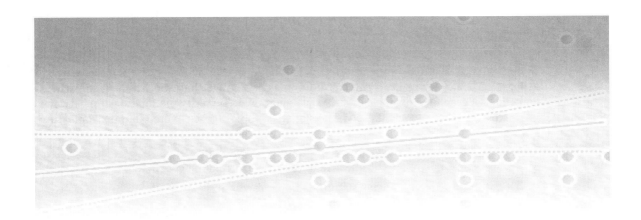

Chapter 4

Dealing with Bivariate Data

4.1 Introduction

When two observations or measurements are made on each member of a sample, we have what is generally termed *bivariate data*. In this chapter, we shall show how to construct informative graphical displays of such data and how to quantify the relationship between the two variables in such data sets. The statistical topics to be covered in this chapter are:

- Plotting the data—Scatterplots
- Assessing strength of linear relationship—Correlation coefficients
- Fitting a line to the data—Simple linear regression

4.2 Example: Heights and Resting Pulse Rates

The data in Table 4.1 show the heights (in centimeters) and resting pulse rates (beats per minute) for a sample of hospital patients. The main question of interest is whether there is any relationship between height and pulse rate.

Table 4.1 Heights (Cm) and Resting Pulse (Bpm) Data

160	68	160	78	185	80	160	80
167	80	170	90	163	95	182	80
162	84	177	80	177	80	168	80
175	80	166	72	165	76	155	80
185	80	170	80	182	100	175	104
162	80	148	82	162	88	168	80
173	92	175	76	172	90	180	68
167	92	160	84	177	90	175	84
170	80	153	70	168	90	145	64
170	80	185	80	178	80	170	84
163	80	165	82	182	76	175	72
158	80	165	84	167	80		
157	80	172	116	170	84		

4.2.1 Plotting Heights and Resting Pulse Rates: The Scatterplot

The height and resting pulse data set is called *bivariate* since two variables are measured for each individual. The separate variables in the data set can, of course, each be summarized and graphed using the methods described in Chapter 2. Of more importance and more interest for bivariate data is to describe and graph the data in a way that lends insights as to how the two variables are related. Let's begin by looking at the most commonly used graphic for bivariate data namely the *scatterplot*—an *xy* plot of the two variables which has been in use since at least the 18th century and has many virtues; indeed, according to Tufte (1983):

> The relational graphic—in its barest form the scatterplot and its variants—is the greatest of all graphical designs. It links at least two variables encouraging and even imploring the viewer to assess the possible causal relationship between the plotted variables. It confronts causal theories that *x* causes *y* with empirical evidence as to the actual relationship between *x* and *y*.

The data giving heights and resting pulse rates shown in Table 4.1 are already available in a SAS data set, **resting**. As in the previous chapter, we will keep the analysis of each data set separate by creating a process flow window for each. So, before adding the data set, we rename the default process flow to **resting**, by right-clicking on the **Process Flow** tab and selecting **Rename**. Then, we add the data set to the process flow.

1. Select **File≻Open≻Data ≻Local Computer**.

2. Browse to the folder containing the SAS data sets (**c:\saseg\sasdata**).

3. Select the file and click **Open**.

For the scatterplot:

1. Select **Graph≻Scatter Plot**.

2. Under **Scatter Plot**, select **2D Scatter Plot**.

3. Under **Task Roles**, drag **height** to **Horizontal** and **pulse** to **Vertical**.

4. Click **Run**.

The resulting plot is shown in Figure 4.1 and suggests that increasing height is generally (although not universally) associated with an increase in resting pulse, and that the relationship between the two variables is, approximately at least, *linear*; i.e., it can be described by a straight line (see Section 4.2.3).

Figure 4.1 Scatterplot of Pulse against Height

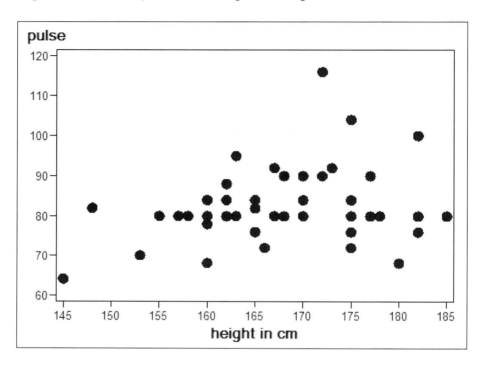

4.2.2 Quantifying the Relationship between Resting Pulse Rate and Height: The Correlation Coefficient

How can we summarize and quantify any relationship between two variables indicated in the scatterplot of the two variables in a single number? What is needed is to measure the *correlation* between the two variables using a *correlation coefficient*. For two continuous variables, we can use *Pearson's correlation coefficient*, also known as the *product-moment correlation coefficient*.

The product moment correlation coefficient is the ratio of the sum of products of differences of each variable from its mean divided by the square roots of the two sums of squares about the mean. Altman (1991) includes more details of how the coefficient is calculated. The product moment coefficient takes values between –1 and 1. Negative values indicate that large values of *x* are associated with small values of *y* and vice versa. Positive values indicate the reverse. The correlation coefficient has a maximum value of

+1 when the points in the scatterplot all lie exactly on a straight line and the variables are positively correlated. The correlation coefficient has a minimum value of –1 when all the points lie exactly on a straight line and the variables are negatively correlated. When the correlation coefficient is zero, the variables are said to be *uncorrelated*.

In essence, the correlation coefficient is a measure of how closely the points in the scatterplot are to a straight line; it measures the *linear relationship* between two variables; nonlinear relationships may be missed or underestimated by it. For example, Figure 4.2 shows a perfect *nonlinear relationship* between two variables for which the correlation coefficient takes the value 0, and Figure 4.3 shows another perfect nonlinear relationship for which the coefficient is not 1. The examples in Figures 4.2 and 4.3 demonstrate the need to use the scatterplot alongside the correlation coefficient when assessing relationships between variables. Use of the correlation coefficient alone is insufficient and can lead to misinterpretation of the data.

Figure 4.2 Perfect Nonlinear Relationship between Two Variables for Which the Correlation Coefficient Is Almost Zero

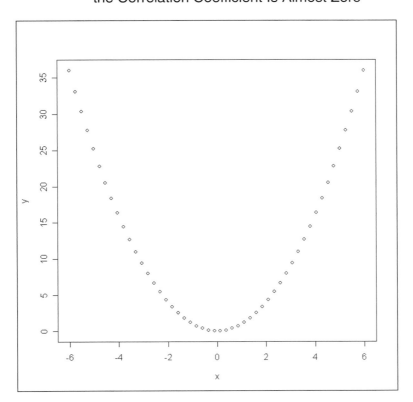

Figure 4.3 Perfect Nonlinear Relationship between Two Variables for Which the Correlation Coefficient Is Nevertheless Not One

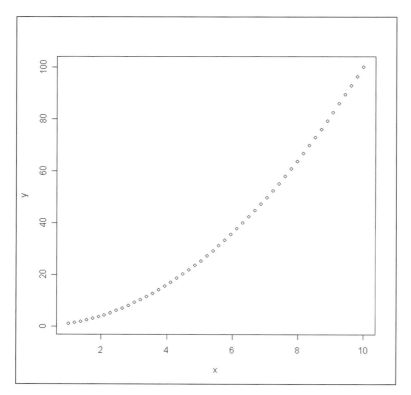

To calculate the product moment correlation coefficients for height and resting pulse rate data:

1. Select the **resting** data set.

2. Select **Analyze➤Multivariate➤Correlations**.

3. Under **Task Roles**, assign both variables to be **Analysis variables**.

4. Click **Run**.

The results are shown in Table 4.2.

Table 4.2 Correlation Coefficients for the Height and Resting Pulse
Rate Data

Pearson Correlation Coefficients, N = 50 Prob > |r| under H0: Rho=0		
	Height	**Pulse**
Height	1.00000	0.21822 0.1279
Pulse	0.21822 0.1279	1.00000

The correlation between height and resting pulse is 0.22 which indicates a relatively weak
positive association between the two variables. A correlation coefficient calculated from
a sample of observations is an *estimate* of the corresponding value in the population (in
the same way that the sample mean is an estimate of the population mean; see Chapter 2).
Consequently, we may want to use the sample correlation as the basis of a test of some
hypothesis about the population correlation. The most common hypothesis of interest is
that the population value is 0; i.e., there is no linear relationship between the two
variables. Under the hypothesis of no linear relationship, a suitable test statistic is

$t = r\sqrt{\dfrac{n-2}{1-r^2}}$ where *n* is the sample size and *r* the sample correlation coefficient. If the

hypothesis of zero population correlation is true, the statistic is known to have a Student's
t distribution with *n*–2 degrees of freedom. The result of the test is labeled
Prob > \|r\| under H0: Rho=0 in the results in Table 4.2. So for height and resting pulse
with a *p*-value of 0.13, there is no evidence that the two variables are related; the
population correlation between the two variables may well be 0.

4.2.3 Heights and Resting Pulse Rates: Simple Linear Regression

Rather than simply measuring the correlation between two variables, we would often like
to derive an equation that links one variable to the other and might, in some situations, be
used for *predicting* the values of one variable from the values of the other. And if such an
equation can be derived, it often is also useful to add it to the scatterplot of the two
variables to highlight their relationship. Most commonly, we wish to find the straight line
that best fits the observed data. Fitting a straight line involves *simple linear regression*

and *least squares estimation*, both of which are described in detail in Altman (1991). But essentially we postulate the following model for the data and then estimate the model's two *parameters* α (the intercept of the line) and β (the slope of the line):

$$y_i = \alpha + \beta x_i + \varepsilon_i$$

In the model above, x_i, y_i represent the observed values of the two variables for the *i*th subject in the sample of observations, and ε_i represents the error; i.e., the amount by which y_i differs from its value as predicted by the model, namely $\alpha + \beta x_i$. The formulae for the sample estimates of α and β are given explicitly in Altman (1991).

We can fit the simple linear regression model to the heights and resting pulse rate data as follows:

1. Select the **resting** data set.

2. Select **Analyze➤Regression➤Linear**.

3. Under **Task Roles**, assign **pulse** the role of **Dependent variable** and **height** the role of **Explanatory variables** (Display 4.1). Note that there is no distinction between **Quantitative** and **Classification** variables. In the Linear Regression task, all explanatory variables are assumed to be quantitative.

4. Click **Run**.

Display 4.1 Task Roles Pane for Linear Regression of Resting Pulse Data

The results are shown in Table 4.3.

Table 4.3 Results of Fitting a Simple Linear Regression Model to the Height and Pulse Rate Data

Number of Observations Read	50
Number of Observations Used	50

Analysis of Variance					
Source	DF	Sum of Squares	Mean Square	F Value	Pr > F
Model	1	186.32129	186.32129	2.40	0.1279
Error	48	3726.17871	77.62872		
Corrected Total	49	3912.50000			

Root MSE	8.81072	R-Square	0.0476
Dependent Mean	82.30000	Adj R-Sq	0.0278
Coeff Var	10.70561		

Parameter Estimates					
Variable	DF	Parameter Estimate	Standard Error	t Value	Pr > |t|
Intercept	1	46.90693	22.87933	2.05	0.0458
height	1	0.20977	0.13540	1.55	0.1279

The first part of Table 4.3 gives an *analysis of variance table* (see Chapter 5) in which the variation in the *y* variable is partitioned into a part due to the fitted model and a part due to the error term in the model. The associated F-test (see Chapter 5) gives a test of the hypothesis that the population value of the slope is 0 ($H_0 : \beta = 0$). Here, the *p*-value associated with the F-test is 0.13 so there is no evidence for a non-zero slope. (Note that the *p*-value is the same as the previously described test for zero correlation between the two variables; the two tests are, of course, equivalent.)

The most important term in the second part of Table 4.4 is *R-square* which is the square of the correlation between the observed values of the response variable, and the values of the response variable predicted by the fitted model. R-square gives the variance in the response variable *y* that is explained by the *x* variable. Here, the R-square value of 0.0476 shows that only about 5% of the variance in pulse rate is accounted for by height. The last section of Table 4.3 gives the estimated intercept and slope for the model. The slope is estimated to be 0.21 which implies that, for every centimeter increase in height, pulse rate increases by 0.21. But since the standard error of the estimated slope is 0.14, the 95% confidence interval for the slope is [–0.07,0.49] which includes the value zero, as we already knew it would from the result of the F-test discussed above.

To add the fitted line and confidence limits for the line to the scatterplot of the two variables, proceed as follows:

1. Reopen the **Linear Regression** task (double-click or right-click **Open**).

2. Under **Plots≻Predicted**, select **Observed vs independents** and **Confidence limits** (Display 4.2).

3. Click **Run**.

4. Replace previous results.

The resulting plot is shown in Figure 4.4. We can see that a horizontal line (i.e., one with slope zero) could easily be fitted between the two confidence limits.

Display 4.2 Selecting Plots of Predicted Values for the Resting Pulse Data

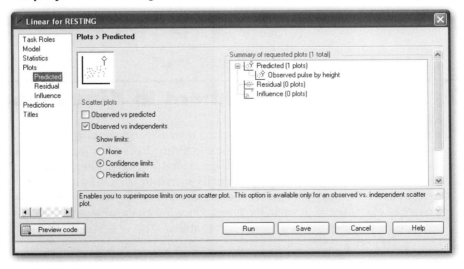

Figure 4.4 Scatterplot of Pulse and Height Data Showing Fitted Linear Regression and Confidence Interval for the Fit

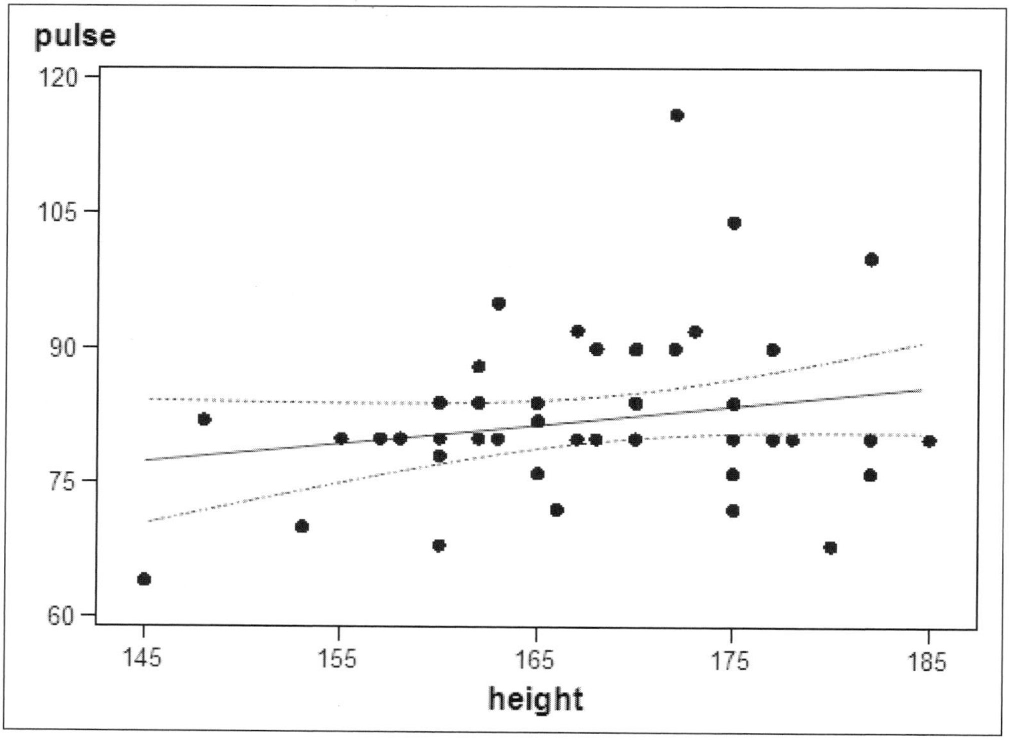

4.3 Example: An Experiment in Kinesiology

The data shown in Table 4.4 were collected in an experiment in *kinesiology* (a natural health care system which uses gentle muscle testing to evaluate many functions of the body in the structural, chemical, neurological, and biochemical realms). A subject performed a standard exercise at a gradually increasing level. Two variables were measured: the first oxygen uptake, and the second expired ventilation which is related to the rate of exchange of gases in the lungs. Once again, the objective is to investigate the relationship between the two measured variables.

Table 4.4 Oxygen Uptake and Expired Ventilation Data

574	21.9	1639	29.2	2766	55.8	3844	100.9
592	18.6	1787	32.0	2812	54.5	3878	103.0
664	18.6	1790	27.9	2893	63.5	4002	113.4
667	19.1	1794	31.0	2957	60.3	4114	111.4
718	19.2	1874	30.7	3052	64.8	4152	119.9
770	16.9	2049	35.4	3151	69.2	4252	127.2
927	18.3	2132	36.1	3161	74.7	4290	126.4
947	17.2	2160	39.1	3266	72.9	4331	135.5
1020	19.0	2292	42.6	3386	80.4	4332	138.9
1096	19.0	2312	39.9	3452	83.0	4390	143.7
1277	18.6	2475	46.2	3521	86.0	4393	144.8
1323	22.8	2489	50.9	3543	88.9		
1330	24.6	2490	46.5	3676	96.8		
1599	24.9	2577	46.3	3741	89.1		

4.3.1 Oxygen Uptake and Expired Ventilation: The Scatterplot

The data on oxygen uptake and expired ventilation, shown in Table 4.2, are also available in a SAS data set, **anaerob.** Before adding them, we create a new process flow window (**File≻New≻Process Flow**) and rename it **anaerob.** Then, we add the data to it in the same way as for the **resting** data set in the previous section. Repeat the scatterplot task assigning **o2in** to the horizontal axis and **airout** to the vertical axis. The result is shown in Figure 4.5, which clearly demonstrates that there is a strong relationship between oxygen uptake and expired ventilation, but that this relationship is distinctly *nonlinear*; as oxygen uptake increases, expired ventilation accelerates making the relationship between the two variables depart from a straight line form.

Figure 4.5 Scatterplot of Oxygen Uptake and Expired Ventilation

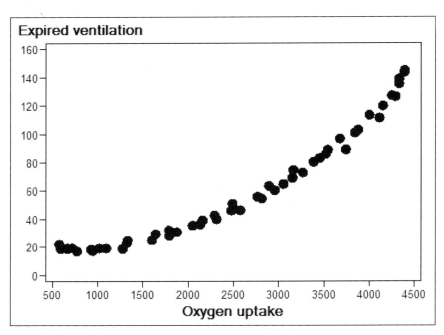

The correlation coefficient for oxygen uptake and expired ventilation can be found in the same way as described in the previous section for height and resting pulse rate. The results are shown in Table 4.5.

Table 4.5 Correlation for Oxygen Uptake and Expired Ventilation

Pearson Correlation Coefficients, N = 53 Prob > |r| under H0: Rho=0		
	o2in	Airout
o2in Oxygen uptake	1.00000	0.95498 <.0001
airout Expired ventilation	0.95498 <.0001	1.00000

For oxygen uptake and expired volume, the correlation is 0.95. Since we know the relationship to be nonlinear, use of the coefficient is not totally informative about the nature of the data and so, although a very small p-value associated with the test of zero correlation provides strong evidence that the two variables are related, the scatterplot indicates that the relationship is not linear. This example emphasizes that it is generally good practice to always have the scatterplot of two variables visible when trying to interpret the correlation coefficient between them.

4.3.2 Expired Ventilation and Oxygen Uptake: Is Simple Linear Regression Appropriate?

We can repeat what was done in the previous subsection for the expired ventilation and oxygen uptake data to give Table 4.6 and Figure 4.6. Here, the test for a zero slope has a very small associated p-value; there is strong evidence that the slope is not zero. The R-square value shows that 91% of the variation in expired ventilation can be attributed to variation in oxygen uptake. But the plot in Figure 4.6 shows that, despite the seemingly impressive statements about the linear fit, it is not the correct model for the ventilation/oxygen data; readers are advised to carry out Exercise 4.3 and fit a more appropriate model for the data.

Table 4.6 Results of Fitting a Linear Regression Model to the Expired Ventilation and Oxygen Uptake Data

Number of Observations Read	53
Number of Observations Used	53

Analysis of Variance					
Source	DF	Sum of Squares	Mean Square	F Value	Pr > F
Model	1	75555	75555	528.40	<.0001
Error	51	7292.38118	142.98787		
Corrected Total	52	82848			

Root MSE	11.95775	R-Square	0.9120
Dependent Mean	60.70755	Adj R-Sq	0.9103
Coeff Var	19.69731		

Parameter Estimates						
Variable	Label	DF	Parameter Estimate	Standard Error	t Value	Pr > |t|
Intercept	Intercept	1	-18.44873	3.81520	-4.84	<.0001
o2in	Oxygen uptake	1	0.03114	0.00135	22.99	<.0001

Figure 4.6 Plot of Expired Ventilation and Oxygen Uptake Data Showing Fitted Linear Regression and Its Confidence Limits

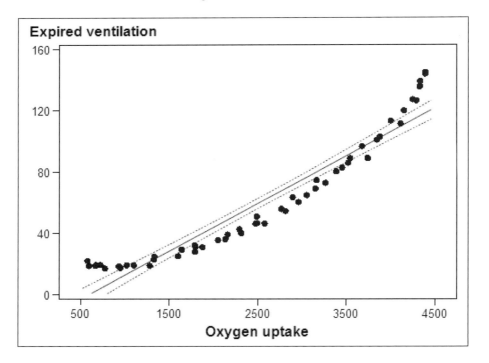

4.4 Example: U.S. Birthrates in the 1940s

The data in Table 4.7 give the monthly U.S. births per thousand population for the years 1940 to 1948. Here, we would like to explore the data for any interesting patterns that may tell a story about the data.

Read along the rows to get monthly observations starting in 1940.

Table 4.7 U.S. Monthly Birthrates between 1940 and 1943

1890	1957	1925	1885	1896	1934	2036	2069	2060
1922	1854	1852	1952	2011	2015	1971	1883	2070
2221	2173	2105	1962	1951	1975	2092	2148	2114
2013	1986	2088	2218	2312	2462	2455	2357	2309
2398	2400	2331	2222	2156	2256	2352	2371	2356
2211	2108	2069	2123	2147	2050	1977	1993	2134
2275	2262	2194	2109	2114	2086	2089	2097	2036
1957	1953	2039	2116	2134	2142	2023	1972	1942
1931	1980	1977	1972	2017	2161	2468	2691	2890
2913	2940	2870	2911	2832	2774	2568	2574	2641
2691	2698	2701	2596	2503	2424			

4.4.1 Plotting the Birthrate Data: The Aspect Ratio of a Scatterplot

An important aspect of a scatterplot that can greatly influence our ability to recognize patterns in the plot is the *aspect ratio*, the physical length of the vertical axis relative to that of the horizontal axis. By default, SAS Enterprise Guide scales plots and other graphics to fill the available graphics area typically resulting in an aspect ratio of around 3:4. To illustrate how changing the aspect ratio of a scatterplot can help understand what the data might be trying to tell us, we shall use the birthrate data given in Table 4.7. First, we create a new process flow window:

1. Select **File≻New≻Process Flow**.

2. Rename it to **usbirths** (right-click on the **Process Flow** tab ≻**Rename**).

The data are in a file **usbirths.dat** in the data directory (**c:\saseg\data**). Import them to the project.

1. Select **File≻Import Data**.

2. Select **Local Computer**, navigate to the location of the file, and click **Open**. There is only a single column of data, so the default import options will suffice.

3. Under **Column Options**, rename the column to **births**.

4. Click **Run**.

The timing of the data is implicit in the order. To create a variable to represent this information:

1. Select **Analyze≻Time Series≻Prepare Time Series Data**.

2. Under **Task Roles,** drag **New TimeId** to **Time ID variable** and rename it **month**.

3. **Monthly** is the default interval, so set the staring date to **1/1/1940**. The screen should now look like Display 4.3.

4. Click **Run**.

Display 4.3 Preparing Time Series of U.S. Births Data: Task Roles Pane

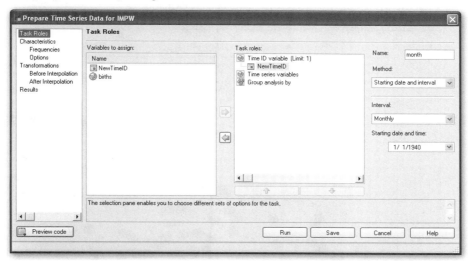

We can construct a scatterplot of the birthrates against month with the default aspect ratio:

1. Select **Graph≻Scatter Plot**.

2. Under **Task Roles**, assign **births** to **Vertical** and **month** to **Horizontal**.

3. Click **Run**.

The resulting plot is shown in Figure 4.7. The plot shows that the U.S. birthrate was increasing between 1940 and 1943, decreasing between 1943 and 1946, rapidly increasing during 1946, and then decreasing again during 1947–1948. As Cook and Weisberg (1982) comment:

> These trends seem to deliver an interesting history lesson since the U.S. involvement in World War II started in 1942 and troops began returning home during the part of 1945, about nine months before the rapid increase in the birthrate.

Figure 4.7 Scatterplot of Birthrate V Month

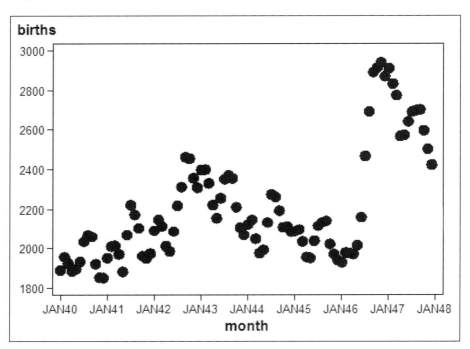

Now let's see what happens when we alter the aspect ratio of the plot:

1. Reopen the **Scatter Plot** task (double-click or right-click **Open**).

2. Under **Appearance≻Chart Area**, select **Specify custom chart size**, and enter **600** for **Width** and **200** for **Height**.

3. Under **Appearance≻Plots**, reduce the size of the **data point marker** to around a third (3 points).

4. Click **Run**.

5. Do not replace the previous results.

The resulting graph appears in Figure 4.8.

Figure 4.8 Scatterplot of Birthrate V Month with Aspect Ratio 0.3

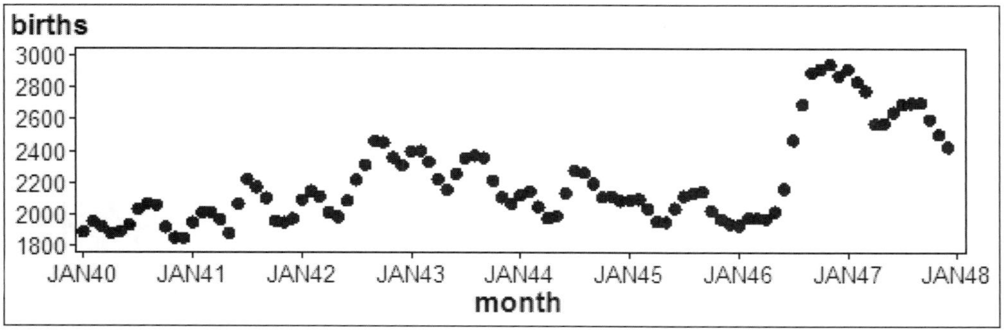

The new plot displays many peaks and troughs and suggests perhaps some minor within-year trends in addition to the global trends apparent in Figure 4.7.

A clearer picture is obtained by plotting only a part of the data; here, we will plot the observations for the years 1940–1943. To do this, we begin by creating a filter to include only data for the years 1940–1943:

1. Select the data set (click on the icon labeled **Modified Time Series data for SASUSER.IMPW**).

2. Select **Data▷Filter and Query**. This opens the Query Builder window.

3. Under the **Filter Data** tab, drag and drop **month**.

4. The Edit Filter window opens. Change **Operator** to **Less than**; click the drop-down button next to the **Value** box, and click **Get Values** in the pop-up window.

5. From the list that appears, select **'1Jan1944'd**.

6. Click **OK** (see Display 4.4).

7. Click **Run**.

Display 4.4 Filter Data Pane: Selecting Data Prior to 1944

Select Data | Filter Data | Sort Data
Filter the raw data

TSDSTIMESERIESOUTDATAIMPW.month < '1Jan1944'd

WHERE TSDSTIMESERIESOUTDATAIMPW.month < '1Jan1944'd

The results are displayed showing only observations for 1940–1943. Select (click) this data set in the Process Flow window and then construct a scatterplot of births by month using the same options as for Figure 4.7. The result is shown in Figure 4.9.

Figure 4.9 Scatterplot of Birthrate V Month for Years 1940–1943; Aspect Ratio=0.3

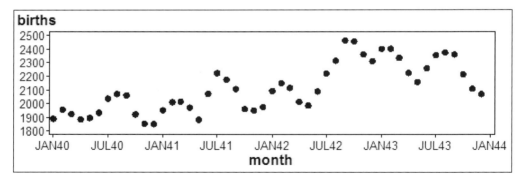

Now, a within-year cycle is clearly apparent with the lowest within-year birthrate at the beginning of the summer and the highest birthrate occurring in the autumn. This pattern can be made clearer in a line plot with the same options as Figure 4.7.

1. Select **Graph≻Line Plot**.

2. Under **Line Plot**, the default, simply labeled **Line Plot**, is the type we want. Double-click **Line Plot**.

3. Under **Task Roles**, assign **births** to **Vertical** and **month** to **Horizontal**.

4. Under **Appearance≻Chart Area**, select **Specify custom chart size**, and enter **600** for **Width** and **200** for **Height**.

5. Click **Run**.

The new plot appears in Figure 4.10.

Figure 4.10 Scatterplot of Birthrate V Month for Years 1940–1943 with Observations Joined and Aspect Ratio=0.3

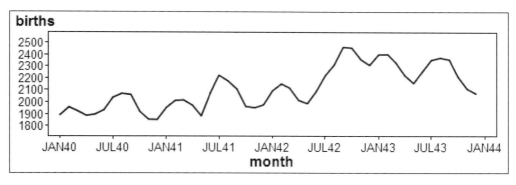

By reducing the aspect ratio to 0.25 and replotting all 96 observations with a line plot, both the within-year and global trends become clearly visible. Select the full data set (click on the icon labeled **Modified Time Series data for SASUSER.IMPW**), then run the **Line Plot** task for **month** and **births** with the **Chart Area** set to **600** by **150**.

The result is shown in Figure 4.11.

Figure 4.11 Scatterplot of Birthrate V Month with Observations Joined and Aspect Ratio 0.25

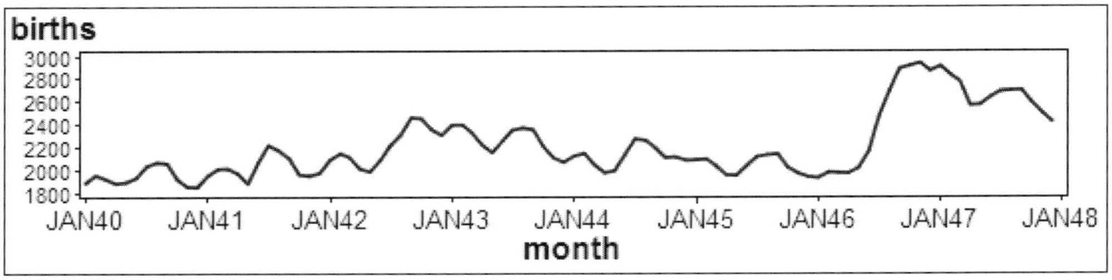

4.5 Exercises

Exercise 4.1 The **mortality** data set contains mortality rates due to malignant melanoma of the skin for white males during the period 1950–1969, for each state on the U.S. mainland. Also given are the latitude and longitude of the center of each state. Construct scatterplots of mortality against latitude and mortality against longitude. In each case, find the corresponding correlation coefficient. Interpret your findings.

Mortality Rates Due to Malignant Melanoma in the U.S.

State	Mortality	Latitude	Longitude
Alabama	219	33.0	87.0
Arizona	160	34.5	112.0
Arkansas	170	35.0	92.5
California	182	37.5	119.5
Colorado	149	39.0	105.5
Connecticut	159	41.8	72.8
Delaware	200	39.0	75.5
Washington DC	177	39.0	77.0
Florida	197	28.0	82.0
Georgia	214	33.0	83.5
Idaho	116	44.5	114.0

(continued)

State	Mortality	Latitude	Longitude
Illinois	124	40.0	89.5
Iowa	128	42.2	93.8
Kansas	166	38.5	98.5
Kentucky	147	37.8	85.0
Louisiana	190	31.2	91.8
Maine	117	45.2	69.0
Maryland	162	39.0	76.5
Massachusetts	143	42.2	71.8
Michigan	117	43.5	84.5
Minnesota	116	46.0	94.5
Mississippi	207	32.8	90.0
Missouri	131	38.5	92.0
Montana	109	47.0	110.5
Nebraska	122	41.5	99.5
Nevada	191	39.0	117.0
New Hampshire	129	43.8	71.5
New Jersey	159	40.2	74.5
New Mexico	141	35.0	106.0
New York	152	43.0	75.5
North Carolina	199	35.5	79.5
North Dakota	115	47.5	100.5
Ohio	131	40.2	82.8
Oklahoma	182	35.5	97.2
Oregon	136	44.0	120.5
Pennsylvania	132	40.8	77.8
Rhode Island	137	41.8	71.5
South Carolina	178	33.8	81.0
South Dakota	86	44.8	100.0
Tennessee	186	36.0	86.2
Texas	229	31.5	98.0

(continued)

State	Mortality	Latitude	Longitude
Utah	142	39.5	111.5
Vermont	153	44.0	72.5
Virginia	166	37.5	78.5
Washington	117	47.5	121.0
West Virginia	136	38.8	80.8
Wisconsin	110	44.5	90.2
Wyoming	134	43.0	107.5

Exercise 4.2 Construct a scatterplot of the data in expired ventilation and oxygen uptake data in Table 4.2. Add the fitted *quadratic* curve to the plot; i.e., a curve of the form

$$y_i = \alpha + \beta_1 x_i + \beta_2 x_i^2 + \varepsilon_i$$

Exercise 4.3 The **index** data set gives the values of a food price index and a house price measure for the U.K. for each year from 1971 to 1989. By constructing suitable plots, investigate how the two price measures change over time, and how the changes are related. (Experiment with changing the aspect ratio of the plots you create.)

Year	Food index	House Index
1971	155.6	60
1972	169.4	79
1973	194.9	107
1974	230.0	113
1975	288.9	124
1976	346.5	134
1977	412.4	148
1978	441.6	177
1979	494.7	227
1980	554.5	272
1981	601.3	280
1982	648.6	285
1983	669.2	317

(continued)

Year	Food index	House Index
1984	706.7	342
1985	728.8	373
1986	752.6	436
1987	775.6	513
1988	802.4	646
1989	847.7	750

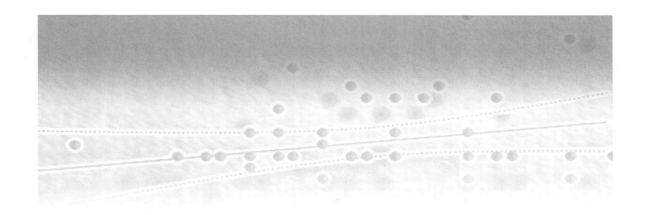

Chapter 5

Analysis of Variance

5.1 Introduction

In this chapter, we will describe how to analyze data in which a response variable of interest is measured in different levels of one or more categorical *factor variables*. The statistical topics to be covered are:

- Analysis of variance for the one way design
- Factorial designs, balanced and unbalanced
- Type I and Type III sums of squares
- Multiple comparison tests

5.2 Example: Teaching Arithmetic

In an experiment to compare different methods of teaching arithmetic (Wetherill 1982), 45 students were divided randomly into five groups of equal size. Two groups—1 and 2—were taught by the current method, and three—3, praised; 4, reproved; 5, ignored—were taught by one of three new methods. At the end of the investigation, all pupils took a standard test with the results shown in Table 5.1. What conclusions can be drawn about possible differences between teaching methods?

Table 5.1 Data on Teaching Methods

Teaching method	Test results
1	17,14,24,20,24,23,16,15,24
2	21,23,13,19,13,19,20,21,16
3	28,30,29,24,27,30,28,28,23
4	19,28,26,26,19,24,24,23,22
5	21,14,13,19,15,15,10,18,20

Before beginning any formal analysis of the data sets in the previous section, it will be useful to consider some summary statistics and graphics for the data since both will be helpful in both gaining informal insights into the data and aiding the interpretation of the formal testing to be described later.

5.2.1 Initial Examination of the Teaching Arithmetic Data with Summary Statistics and Box Plots

The teaching data in Table 5.1 are already available in a SAS data set, **teaching**. Add the data set to the project:

1. Select **File≻Open≻Data≻Local Computer**.

2. Browse to the folder containing the SAS data set (**c:\saseg\sasdata**).

3. Select the file.

4. Click **Open**.

We saw in Chapter 3 how to produce tables of counts and percentages. Here, we need tables containing other summary statistics, specifically *means* and *standard deviations*. To do this:

1. Select **Describe≻Summary Tables**.

2. Under **Task Roles**, *classification variables* are those whose values will be used to form the rows and/or columns of the table, and *analysis variables* are those whose values are to be summarized within the table. In this case, **method** is a **Classification variable** and **result** is an **Analysis variable**.

3. Under **Summary Tables**, tables are constructed by dragging variables across to the appropriate position in the **Preview** pane. If a variable is not appropriate for the position it is dragged to, the cursor will change to a stop sign (⊘).

4. Drag **method** to the row position (left side). A column headed **N** appears. This is because the default summary statistic for **Classification variables** is a **count**.

5. Drag **result** to the column position on either edge of the box containing **N**. **Sum** appears below it, as this is the default summary statistic for an **Analysis variable**.

6. From the **Available statistics** pane, drag **mean** to the left edge of the **N** box.

7. Repeat Step 6 with **stdev** (standard deviation).

8. Remove **Sum** by dragging it off the **Preview** pane.

9. Among the **Available variables** there is an additional variable, **All**. This is used to form totals. Drag it to the bottom edge of the **method** box. The **Summary Tables** pane should now look like the example in Display 5.1.

10. Click **Run**.

The results are shown in Table 5.2.

Display 5.1 Summary Tables Pane for the Teaching Data

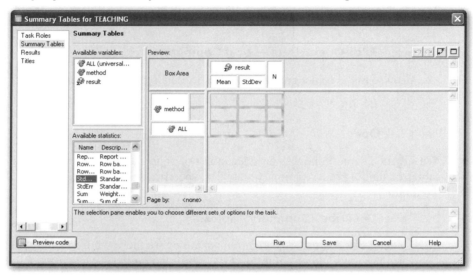

Table 5.2 Summary Statistics for Teaching Methods Data

	result		
	Mean	**StdDev**	**N**
Method			
1	19.67	4.21	9
2	18.33	3.57	9
3	27.44	2.46	9
4	23.44	3.09	9
5	16.11	3.62	9
All	21.00	5.21	45

A useful graphic for these data consists of the box plots of the observations made under each teaching method.

1. Select **Graph≻Box Plot**.

2. Under **Task Roles**, assign **method** to **Horizontal** and **result** to **Vertical.**

3. Click **Run.**

The resulting box plots are shown in Figure 5.1. The box plot and the summary statistics in Table 5.2 suggest some interesting differences between the five methods. Method 3 (praised), for example, appears to give far better results than the other methods although there are two distinct outliers that perform considerably less well than the other students taught by method 3. The observations for teaching method 5 are quite skewed.

Figure 5.1 Box Plots of the Five Teaching Methods in Table 5.1

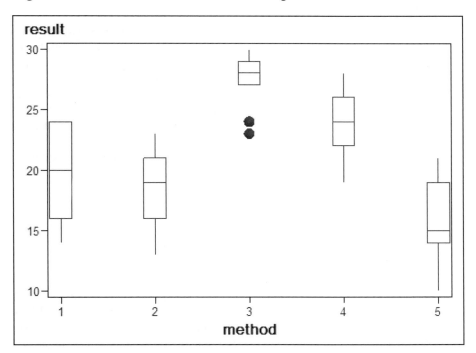

5.2.2 Teaching Arithmetic: Are Some Teaching Methods for Teaching Arithmetic Better Than Others?

The teaching arithmetic study is an example of what is generally known as a *one-way design*. In such designs, interest centers on assessing the effect of a single factor variable (teaching method here) on a response variable (test score). The question posed in such a design is "Do the populations corresponding to the different levels of the factor variable have different means?" Consequently, the null hypothesis that we aim to test is the equality of means of the populations; i.e.:

$$H_0 : \mu_1 = \mu_2 = \ldots\ldots = \mu_k$$

where $\mu_1, \mu_2, \ldots.\mu_k$ are the population means and k is the number of levels of the factor variable.

In Chapter 2, we described how to use Student's *t*-test to test the equality of *two* population means, and here we might post the question, "Why not simply apply the test to each pair of means in our one-way design to assess the null hypothesis above?" The reason that such an approach is inappropriate is that each of the $N=k(k-1)/2$ *t*-tests we would perform is tested at the usual 5% significance level; the probability of rejecting the equality of at least one pair of population means when the null hypothesis is true (P) is greater than the nominal significance level of 0.05 when k is three or larger; details of the involved calculations that demonstrate that this is so are given in Everitt (1996). Here, we simply give some numerical results that illustrate the problem with the *t*-test approach:

k	N	P
3	3	0.14
4	6	0.26
10	45	0.90

The appropriate approach to the analysis of data arising from a one-way design is the *analysis of variance* (ANOVA), the phrase having been coined by Ronald Aylmer Fisher. Fisher defined it as "the separation of variance ascribable to one group of causes from the variance ascribable to the other groups." Stated another way, the analysis of variance is a partitioning of the total variance in a set of data into a number of component parts. In a one-way design, for example, we separate the total variance into two parts: a part due to differences in the sample means of the levels of the factor variable (*between groups variance*) and a part measuring variance within the levels of the factor variable (*within groups variance*).

If the null hypothesis of equality of means is correct, then both the between groups' and the within groups' variances are estimating the same population quantity; if the null hypothesis is wrong, then the between groups' variance is estimating a *larger* population quantity than the within group variance. Consequently, a test of the equality of the two population variances (between groups and within groups)—based on the two estimates of them—will be a test of the null hypothesis about the population means that we are interested in. The appropriate test for the equality of two variances is what is known as an *F-test*. Full details of the analysis of variance for a one-way design are given in Everitt (1996).

To apply the analysis of variance to the teaching method data in Table 5.1:

1. Select **Analyze≻ANOVA≻One-Way ANOVA**.

2. Under **Task Roles**, make **result** the **Dependent variable** and **method** the **Independent variable**.

3. Click **Run**.

The results are shown in Table 5.3. In this table, the part that is of most importance for us is that which gives the result of the partition of the variation in the data, since this is where we find the result of the F-test for assessing the equality of means hypothesis. We see that the *p*-value associated with the F-test is very small (<0.0001), so there is considerable evidence that the teaching methods do indeed differ with respect to their mean arithmetic test scores.

Table 5.3 Analysis of Variance Results for Teaching Methods Data

Class Level Information		
Class	Levels	Values
Method	5	1 2 3 4 5

Number of Observations Read	45
Number of Observations Used	45

Source	DF	Sum of Squares	Mean Square	F Value	Pr > F
Model	4	722.666667	180.666667	15.27	<.0001
Error	40	473.333333	11.833333		
Corrected Total	44	1196.000000			

R-Square	Coeff Var	Root MSE	result Mean
0.604236	16.38077	3.439961	21.00000

Source	DF	Anova SS	Mean Square	F Value	Pr > F
Method	4	722.6666667	180.6666667	15.27	<.0001

Assumptions Made by the F-Test

The data collected from a one-way design need to satisfy the following assumptions to make the involved F-test strictly valid:

- The observations in each level of the factor variable arise from a population with a normal distribution.

- The population variances of the different levels of the factor are the same.

- The observations are independent of one another.

The assumptions are often difficult to check, particularly when the number of observations in each group is small. Fortunately, the F-test is known to be relatively robust against departures from both normality and homogeneity of variance, especially when the number of observations in each group is equal or approximately equal. In some cases, a *transformation* of the data—for example, taking logs—may aid in achieving both a normal distribution and homogeneity although interpretation may become more problematical. See Exercise 5.3 for an example of applying a transformation and Everitt (1996) for more details of transformations.

Multiple Comparison Tests: Scheffe's Test

Having shown that there is strong evidence of the effect of teaching method on test score, we may wish to investigate in more detail which methods differ (an overall significant F does *not* imply that *all* means differ). For the required, more detailed examination, we can use one of a variety of what are termed *multiple comparison tests*. Such tests compare each pair of means in turn but take steps to avoid the problem of inflating the type I error discussed earlier in the chapter. Here, we shall use *Scheffe's method*, which is described in detail in Everitt (1996) and which is particularly useful when a large number of comparisons have to be made. To apply Scheffe's method to the teaching methods data:

1. Reopen the **One-Way ANOVA** task (double-click or right-click **Open**).

2. Under **Means≻Comparison**, select **Scheffe's multiple comparison procedure**.

3. Click **Run.**

4. Replace the results of the previous run.

The results are shown in Table 5.4. We see that method 3 differs from methods 1, 2, and 5 but not method 4. Method 5 differs from methods 4 and 3 but not from methods 1 and 2. Methods 4, 1, and 2 do not differ from each other. Essentially, the result from Scheffe's test produces a grouping of the methods into (3, 4), (4, 1, 2), (2, 5).

Table 5.4 Results from Scheffe's Multiple Comparison Procedure Applied to the Teaching Methods Data

Alpha	0.05
Error Degrees of Freedom	40
Error Mean Square	11.83333
Critical Value of F	2.60597
Minimum Significant Difference	5.2356

	Means with the same letter are not significantly different.			
Scheffe Grouping		**Mean**	**N**	**method**
	A	27.444	9	3
	A			
B	A	23.444	9	4
B				
B	C	19.667	9	1
B	C			
B	C	18.333	9	2
	C			
	C	16.111	9	5

5.3 Example: Weight Gain in Rats

The data shown in Table 5.5 come from an experiment to study the gain in weight of rats fed on four different diets, distinguished by amount of protein (low and high) and by source of protein (beef and cereal). Ten rats were randomized to each of the four possible diets. The question of interest is how diet affects weight gain.

Table 5.5 Rat Weight Gain for Diets Differing by the Amount of Protein and Source of Protein

Beef		Cereal	
Low	High	Low	high
90	73	107	98
76	102	95	74
90	118	97	56
64	104	80	111
86	81	98	95
51	107	74	88
72	100	74	82
90	87	67	77
95	117	89	86
78	111	58	92

5.3.1 A First Look at the Rat Weight Gain Data Using Box Plots and Numerical Summaries

For the data on weight gain in rats given in Table 5.5, we first open a new process flow to keep the analyses separate:

1. Select **File➤New➤Process Flow**.

2. To rename the process flow, right-click on the **Process Flow** tab, select **Rename**, and type **weightgain.**

The data are in a SAS data set **weight**. Add the data set to the process flow:

1. Select **File≻Open≻Data≻Local Computer**.

2. Browse to the folder containing the SAS data set (**c:\saseg\sasdata**).

3. Select the file.

4. Click **Open**.

The data set contains four variables: **weightgain**, **source**, **level**, and **diet**. The last variable combines information on the source of protein and the level. This can be used to produce a box plot with all four diet types:

1. Select **Graph≻Box Plot**.

2. Under **Task Roles**, assign **diet** to **Horizontal** and **weightgain** to **Vertical**.

3. Click **Run.**

The result is shown in Figure 5.2.

Figure 5.2 Box Plots for Weight Gain in Rats Data

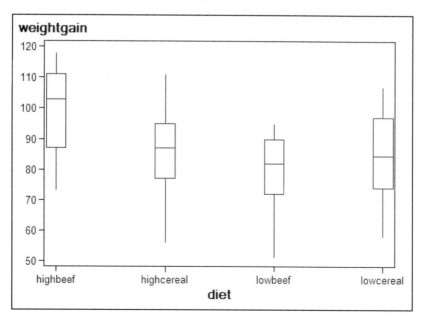

Again, it is also useful to have some numerical summaries for these data. Here, a 2 x 2 table, which was formed from the two levels of source of protein and the two levels of amount of protein showing the corresponding mean and standard deviation, is useful:

1. Select **Describe≻Summary Tables**.

2. Under **Task Roles**, **source** and **level** are assigned as **Classification variables** and **weightgain** is the **Analysis variable**.

3. Under **Summary Tables**, drag **level** to the row position.

4. Drag **weightgain** to the column position.

5. Drag **source** to the top edge of the **weightgain** box.

6. Drag **mean** to the left edge of the **N** box.

7. Drag **stdev** to the right edge of the **mean** box.

8. Remove **N** by dragging it off the **Preview** pane.

9. Click **Run.**

The **Summary Table** pane should look like Display 5.2.

Display 5.2 Summary Table Pane for Weight Gain Data Showing Table Preview

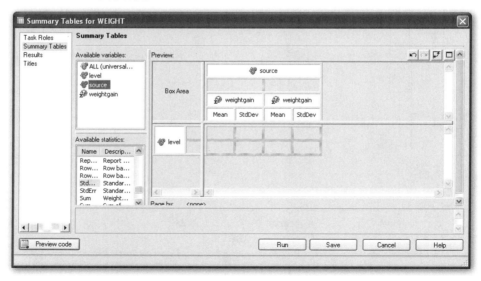

The results are shown in Table 5.6. The standard deviations in each cell are seen to be very similar to each other, a finding which has implications for the formal analysis of the data (see below). The mean weight gain for beef/high is considerably larger than the other three means, which are quite close to each other.

Table 5.6 Numerical Summary Statistics for Rat Weight Gain Data

Crosstabular Summary Report	source			
	beef weightgain		cereal weightgain	
	Mean	StdDev	Mean	StdDev
level				
high	100.00	15.14	85.90	15.02
low	79.20	13.89	83.90	15.71

5.3.2 Weight Gain in Rats: Do Rats Gain More Weight on a Particular Diet?

The data in Table 5.2 are from a simple example of what is known as a *factorial design;* a factorial design involves the simultaneous study of the effect of two or more factor variables on a response variable of interest. In the rats example, the two factors are source and amount of protein given to the rats. As in the previous example involving different teaching methods, questions of interest about these data concern the equality of weight gain for the two levels of source of protein and for the two levels of amount of protein. So, why not apply a one-way analysis of variance to each factor separately? (In our particular example, there are only two levels, so a one-way analysis of variance is equivalent to the *t*-test covered in Chapter 2.) The answer to this question is that such an approach would omit an aspect of a factorial design that is often very important, namely testing whether there is an *interaction* between the two factors. In simple terms, such an effect arises when the effect of applying both factors is either larger (or smaller) than the sum of the effects associated with applying each factor separately. The analysis of variance for a factorial design will include a test for such possible interaction effects. Everitt (1996) includes full details of the analysis of variance for factorial designs.

To apply the analysis of variance to the rat weight gain data, use the **Linear Models** task:

1. Select the **weight** data set.

2. Select **Analyze≻ANOVA≻Linear Models**.

3. Under **Task Roles**, assign **weightgain** as the **Dependent variable** and **source** and **level** as **Classification variables**.

4. Under **Model**, to set up a factorial model including both main effects and their interaction, select both **source** and **level** (CTRL-click on both). The **Main**, **Cross**, and **Factorial** buttons all become active. The **Main** and **Cross** buttons could be used to include the main effects and interaction (**Cross**) in the model separately, but the **Factorial** button does both. Clicking on the **Factorial** button results in **level**, **source**, and **level*source** being inserted in the **Effects** pane.

5. In the **Model Options** pane, deselect **Confidence limits for parameter estimates**.

6. Click **Run**.

The results from running that model are shown in Table 5.7.

Table 5.7 Analysis of Variance Results for the Weight Gain in Rats Data

Class Level Information		
Class	**Levels**	**Values**
Level	2	high low
Source	2	beef cereal

Number of Observations Read	40
Number of Observations Used	40

Source	DF	Sum of Squares	Mean Square	F Value	Pr > F
Model	3	2404.10000	801.36667	3.58	0.0230
Error	36	8049.40000	223.59444		
Corrected Total	39	10453.50000			

R-Square	Coeff Var	Root MSE	Weightgain Mean
0.229980	17.13819	14.95307	87.25000

Source	DF	Type I SS	Mean Square	F Value	Pr > F
Level	1	1299.600000	1299.600000	5.81	0.0211
Source	1	220.900000	220.900000	0.99	0.3269
level*source	1	883.600000	883.600000	3.95	0.0545

Source	DF	Type III SS	Mean Square	F Value	Pr > F
Level	1	1299.600000	1299.600000	5.81	0.0211
Source	1	220.900000	220.900000	0.99	0.3269
level*source	1	883.600000	883.600000	3.95	0.0545

The analysis of variance table in Table 5.7 shows the results of partitioning the variation in the weight gain observations into parts due to amount of protein, to source of protein, and to the interaction of amount and source. The corresponding F-tests show that there is evidence of a difference in weight gain for low and high levels of protein, but no evidence of a difference for source of protein. The F-test for the interaction of the two factors just fails to reach significance at the conventional 5% level, but it may still be of interest to examine in more detail just what such an interaction, if it exists, implies. The test for interaction assesses whether or not the difference between mean weight gain for, say, beef and cereal protein when given at the low level is the same as the corresponding difference when given at the high level. Here, there is some relatively weak evidence that the two differences are not equal. To see more clearly what is happening, we can construct what is sometimes called an *interaction plot*.

Interaction Plots

The interaction plot is essentially a plot of the four cell means and is an option available within the **Linear Models** task.

1. Reopen the **Linear Models** task (double-click or right-click **Open**).

2. Under **Plots≻Means**, select **Dependent means for two-way effects** and **Observed means** (Display 5.3).

3. Click **Run.**

4. Replace the results of the previous run.

Display 5.3 Selecting the Interaction Plot for the Weight Gain in Rats Data

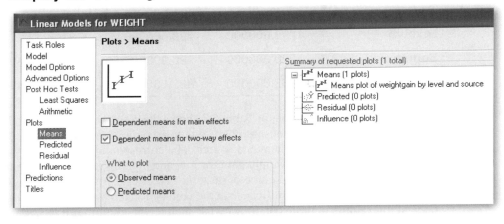

The result is shown in Figure 5.3. The plot suggests that the difference in weight gain for beef and cereal protein is greater when given at the high level than when given at the low level although the effect does not reach the conventional 5% significance level.

Figure 5.3 Interaction Plot for Rat Weight Gain Data

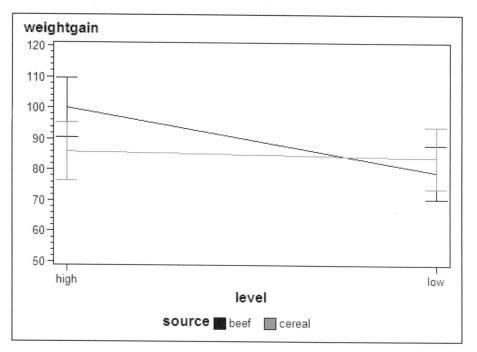

In Table 5.7, there are two analysis of variance tables which are identical except in the labeling of the sums of squares (SS) column where one is labeled *Type I SS* and the other *Type III SS*. In the weight gain in rats example where there are an equal number of observations in each cell of the factorial design, the Type I and Type III methods of computing sums of squares give the same results. However, in an unbalanced design where the number of observations in the cells differ, the use of the Type I and Type III of computing sums of squares give different results, as we will illustrate in the next section.

In a factorial design, the assumptions needed for the F-tests in the analysis of variance table to be strictly valid are similar to the assumptions needed for the one-way design listed earlier, namely, *normality* and *homogeneity*. The homogeneity assumption at least seems appropriate for the rat weight gain data given that we found in Table 5.6 the standard deviations of the observations in each of the four cells of the design are approximately equal.

5.4 Example: Mother's Post-Natal Depression and Child's IQ

The data shown in Table 5.8 were obtained from an investigation into the effect of mothers' post-natal depression on child development. Mothers who gave birth to their first-born child in a major teaching hospital in London were divided into two groups, depressed or not depressed, on the basis of their mental state three months after the birth. The children's fathers were also divided into two groups, namely fathers who had a history of psychiatric illness and fathers who did not. The dependent variable in the study was the child's IQ at age 4 years. Only girl babies were involved in the study. The question of interest here is how post-natal depression affects a child's cognitive development.

Table 5.8 Data Obtained in a Study of the Effect of Post-Natal Depression of the Mother on the Child's Cognitive Development

Mother's depression	Father's history	Child's IQ at 4 years
1	0	103
1	0	124
1	0	124
1	0	104
1	0	96
2	0	92
1	0	125
1	0	99
1	0	103
1	1	98
1	1	101
1	0	104
2	1	97
2	1	95
1	0	120
2	0	105
1	0	124

(continued)

Table 5.8 (*continued*)

Mother's depression	Father's history	Child's IQ at 4 years
1	0	123
1	0	115
1	1	110
1	1	112
1	1	120
1	1	123
2	0	98
1	0	104
2	0	97
1	0	125
1	0	123
1	1	101
2	1	99
1	0	120
2	0	101
1	0	118
1	0	120
1	1	99

Mother's depression: 1=no, 2=yes.
Father's history: 0=no previous psychiatric history, 1=has a previous psychiatric history.

5.4.1 Summarizing the Post-Natal Depression Data

The summary statistics and count of numbers of observations in each cell of the design for the post-natal depression data are obtained in the same way as in the previous examples. Again, we open a new process flow window:

1. Select **File≻New≻Process Flow**.

2. Rename the process flow **Post-natal depression** (right-click on the **Process Flow** tab and select **Rename**).

The data are in a tab-delimited file, **depressionIQ.tab**, with the variable names in the first row. To import the data:

1. Select **File≻Import Data≻Local Computer**, browse to the folder **c:\saseg\data**, select the file and click **Open**.

2. Under **Region to import**, check **Specify line to use as column headings**; line 1 is the default.

3. Under **Text format**, select **Delimited** and **Tab**.

4. Under **Column Options**, check that the variables have been correctly recognized.

5. Click **Run**.

For the summary table:

1. Select **Describe≻Summary Tables**.

2. Under **Task Roles**, assign **childIQ** as the **Analysis variable** and **Mo_depression** and **Pa_history** as the **Classification variables**.

3. Under **Summary Tables**, drag **Pa_history** to the rows position.

4. Drag **childIQ** to the column heading position.

5. Drag **Mo_depression** to the top edge of the **childIQ** box.

6. Drag **mean** to the left edge of the **N** box.

7. Drag **stdev** (**std...**) to the left edge of the **N** box. The **Summary Tables** pane should now look like Display 5.4.

8. Click **Run**.

The results are shown in Table 5.9.

Display 5.4 Summary Tables Pane for Post-Natal Depression Data

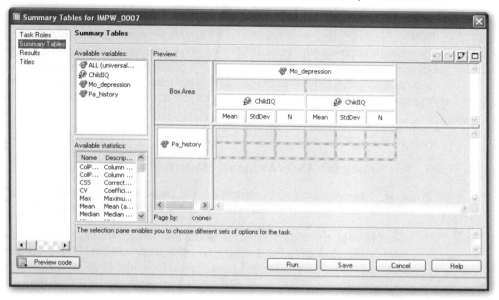

Table 5.9 Numerical Summary Statistics and Cell Counts for IQ Scores from Post-Natal Depression Study

Crosstabular Summary Report	Mo_depression					
	1			**2**		
	ChildIQ			**ChildIQ**		
	Mean	**StdDev**	**N**	**Mean**	**StdDev**	**N**
Pa_history						
0	114.42	10.32	19	98.60	4.83	5
1	108.00	9.77	8	97.00	2.00	3

Table 5.6 suggests that the average IQ is less for children whose mothers suffered from post-natal depression and for children whose father had a previous psychiatric history. Notice also that the numbers of observations in each of the four cells of the table are *not* the same; the design is said to be *unbalanced*.

5.4.2 How Is a Child's IQ Affected by Post-Natal Depression in the Mother?

The data on post-natal depression and IQ have a very similar structure to the data on weight gain in rats: both data sets involve two factor variables and a response variable. But the numbers of observations in each cell of the post-natal depression data are not equal as they are for the weight gain data. The unequal cell size in the post-natal depression data has serious implications for the data analysis as we will see later.

Finding the analysis of variance table for the IQ scores from the post-natal depression study is straightforward.

1. Select **Analyze≻ANOVA≻Linear Models**.

2. Under **Task Roles**, assign **childIQ** as the **Dependent variable** and **Pa_history** and **Mo_depression**, in that order, as the **Classification variables**.

3. Under **Model**, to set up a factorial model including both main effects and their interaction, select both **Pa_history** and **Mo_depression** (CTRL-click on both), and click on **Factorial**. The **Model** pane should look like Display 5.5.

4. Click **Run**.

The results are shown in Table 5.10.

Display 5.5 Model Pane for Post-Natal Depression Data

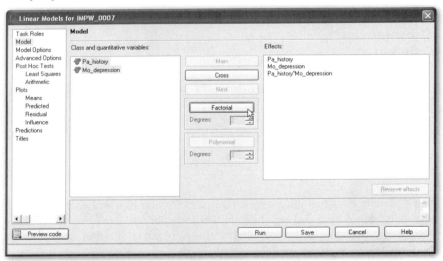

Table 5.10 Analysis of Variance Results for the Post-Natal Depression Study

Class Level Information		
Class	**Levels**	**Values**
Pa_history	2	0 1
Mo_depression	2	1 2

Number of Observations Read	35
Number of Observations Used	35

Source	DF	Sum of Squares	Mean Square	F Value	Pr > F
Model	3	1537.768421	512.589474	5.92	0.0026
Error	31	2685.831579	86.639728		
Corrected Total	34	4223.600000			

R-Square	Coeff Var	Root MSE	ChildIQ Mean
0.364090	8.523852	9.308046	109.2000

Source	DF	Type I SS	Mean Square	F Value	Pr > F
Pa_history	1	282.975000	282.975000	3.27	0.0804
Mo_depression	1	1222.101866	1222.101866	14.11	0.0007
Pa_histor*Mo_depress	1	32.691555	32.691555	0.38	0.5435

Source	DF	Type III SS	Mean Square	F Value	Pr > F
Pa_history	1	90.492912	90.492912	1.04	0.3147
Mo_depression	1	1011.820488	1011.820488	11.68	0.0018
Pa_histor*Mo_depress	1	32.691555	32.691555	0.38	0.5435

In Table 5.10, we see that the Type I and Type III sums of squares for the main effects of a father's psychiatric history and a mother's post-natal depression are *not* the same but they are the same for the interaction term. In a factorial design where there are unequal numbers of observations in each cell of the design, it is no longer possible to partition the variation in the data into *non-overlapping* or *orthogonal* sums of squares representing main effects and interactions as it is in a design with equal numbers of observations in each cell. In an unbalanced two-way layout (for example) with factors A and B, there is a proportion of the variance of the response variable that can be attributed to either A or B. The consequence is that A and B together explain less of the variation of the dependent variable than the sum of which each explains alone. The result is that the sum of squares corresponding to a factor depends on which other terms are currently in the model for the observations, so the sums of squares depend on the order in which the factors are considered and represent a comparison of models. For example, for the order A, B, AxB, the sums of squares are such that:

- **SSA** (sum of squares for A): compares model containing only the A main effect with one containing only the overall mean
- **SSB|A** (sum of squares for B given A is already in the model): compares model with both main effects, but no interaction, with one including only the main effect of A
- **SSAB|A,B** (sum of squares for AxB given that A and B are already in the model): compares model including an interaction and main effects with one including only main effects

These are what are called Type I sums of squares. In contrast, Type III sums of squares represent the contribution of each term to a model including all other possible terms. Thus for two-factor designs, the Type III sums of squares represent:

- **SSA: SSA|B,AB** (sum of squares for A given that B and AxB are in the model)
- **SSB: SSB|A,AB** (sum of squares for B given that A and AxB are in the model)
- **SSAB: SSAB|A,B** (sum of squares for AxB given that A and B are in the model)

SAS also has a Type IV sum of squares, which is the same as Type III unless the design contains empty cells.

In a balanced design, Type I and Type III sums of squares are equal; for an unbalanced design they are not equal, and there have been numerous discussions over which type is most appropriate for the analysis of such designs. Authors such as Maxwell and Delaney (1990) and Howell (1992) strongly recommend the use of Type III sums of squares, and they are the default in SAS Enterprise Guide. Nelder (1977) and Aitkin (1978), however, are strongly critical of "correcting" main effects sums of squares for an interaction term involving the corresponding main effect; their criticisms are based on both theoretical and pragmatic grounds. The arguments are relatively subtle but in essence go something like the following:

- When fitting models to data, the principle of *parsimony* is of critical importance. In choosing among possible models, we do not adopt complex models for which there is no empirical evidence.

- So if there is no convincing evidence of an AB interaction, we do not retain the term in the model. Thus, additivity of A and B is assumed unless there is convincing evidence to the contrary.

- So the argument proceeds that Type III sum of squares for A in which it is adjusted for AB makes no sense.

- First if the interaction term is necessary in the model, then the experimenter will usually wish to consider simple effects of A at each level of B separately. A test of the hypothesis of no A main effect would not usually be carried out if the AB interaction is significant.

- If the AB interaction is not significant, then adjusting for it is of no interest and causes a substantial loss of power in testing the A and B main effects.

The issue does not arise so clearly in the balanced case, for there the sum of squares for a main effect is independent of whether interaction is assumed or not. Thus in deciding on possible models for the data, the interaction term is not included unless it has been shown to be necessary, in which case tests on main effects involved in the interaction are not carried out or, if the tests are carried out, no attempt should be made to interpret them.

The arguments of Nelder and Aitkin against the use of Type III sums of squares are powerful and persuasive. Their recommendation to use Type I sums of squares—combined with doing a number of analyses in which main effects are considered in a number of orders—as the most suitable way in which to identify a suitable model for a data set is also convincing and strongly endorsed by the authors of this book.

So for the post-natal depression data, we will now produce another analysis of variance table in which the main effects are considered in the order of mother's depression followed by father's psychiatric history, which is the reverse order to what was produced in Table 5.10. To do this:

1. Reopen the **Linear Models** task (double-click or right-click **Open**).

2. Under **Model**, select all the effects and click **Remove effects**.

3. To reenter the main effects, select **Mo_depression** and click **Main**.

4. Repeat with **Pa_history**.

5. To reenter the interaction, select both effects and click **Cross**.

6. Click **Run**.

7. Do not replace the results from the previous run.

The results are shown in Table 5.11. Comparing the analysis of variance table in Table 5.11 with that in Table 5.10, we see that the interaction Type I sum of squares are the same but that the main effects sums of squares are different for the two ways of ordering the effects. (The Type III sums of squares are, of course, the same in both Table 5.10 and Table 5.11.) But the conclusion to be made from both analyses is that there is no evidence of an interaction effect and no evidence that father's psychiatric history affects the child's IQ. But there is strong evidence that the child's IQ is associated with the mother's post-natal depression with the occurrence of post-natal depression in the mother appearing to lead to a lower IQ for the child at age four.

Table 5.11 Analysis of Variance Results for the Post-Natal Depression Study after Reordering of Main Effects

Class Level Information		
Class	**Levels**	**Values**
Pa_history	2	0 1
Mo_depression	2	1 2

Number of Observations Read	35
Number of Observations Used	35

Source	DF	Sum of Squares	Mean Square	F Value	Pr > F
Model	3	1537.768421	512.589474	5.92	0.0026
Error	31	2685.831579	86.639728		
Corrected Total	34	4223.600000			

R-Square	Coeff Var	Root MSE	ChildIQ Mean
0.364090	8.523852	9.308046	109.2000

Source	DF	Type I SS	Mean Square	F Value	Pr > F
Mo_depression	1	1300.859259	1300.859259	15.01	0.0005
Pa_history	1	204.217607	204.217607	2.36	0.1349
Pa_histor*Mo_depress	1	32.691555	32.691555	0.38	0.5435

Source	DF	Type III SS	Mean Square	F Value	Pr > F
Mo_depression	1	1011.820488	1011.820488	11.68	0.0018
Pa_history	1	90.492912	90.492912	1.04	0.3147
Pa_histor*Mo_depress	1	32.691555	32.691555	0.38	0.5435

5.5 Exercises

Exercise 5.1 The **rats** data set derives from a study in which the effects of three different poisons and four different treatments on the survival times of rats in hours were of interest. Carry out an appropriate analysis of variance of these data paying particular attention to possible violations of distributional assumptions.

Treatment

Poison	A	B	C	D
1	0.31	0.82	0.43	0.45
	0.45	1.10	0.45	0.71
	0.46	0.88	0.63	0.66
	0.43	0.72	0.76	0.62
2	0.36	0.92	0.44	0.56
	0.29	0.61	0.35	1.02
	0.40	0.49	0.31	0.71
	0.23	1.24	0.40	0.38
3	0.22	0.30	0.23	0.30
	0.21	0.37	0.25	0.36
	0.18	0.38	0.24	0.31
	0.23	0.29	0.22	0.22

Exercise 5.2 The **knee** data set comes from an investigation described by Kapor (1981) in which the effect of knee-joint angle on the efficiency of cycling was studied. Efficiency was measured in terms of distance (km) pedaled on an ergocycle until exhaustion. The experimenter selected three knee-joint angles of particular interest, 50, 70, and 90 degrees. Ten subjects were randomly allocated to each angle. The drag of the ergocycle was kept constant at 14.7N and subjects were instructed to pedal at a constant speed of 20 km/h until exhaustion.

1. Carry out an initial data analysis to assess whether there are any aspects of the data that might be a cause for concern in later analyses.

2. Calculate the appropriate analysis of variance table for the data.

3. Use a variety of multiple comparison tests to explore differences between the population means for each angle and compare their results.

50	70	90
8.4	10.6	3.2
7.0	7.5	4.2
3.0	5.1	3.1
8.0	5.6	6.9
7.8	10.2	7.2
3.3	11.0	3.5
4.3	6.8	3.1
3.6	9.4	4.5
8.0	10.4	3.8
6.8	8.8	3.6

Exercise 5.3 The data in the SAS data set **hypertension** are from a study described by Maxwell and Delaney (1990) in which the effects of three possible treatments for hypertension were investigated. The details of the treatments are as follows:

Treatment	Description	Levels
drug	medication	drug X, drug Y, drug Z
biofeed	psychological feedback	present, absent
diet	special diet	present, absent

All 12 combinations of the three treatments were included in a 3×2×2 design. Seventy-two subjects suffering from hypertension were recruited to the study with six being randomly allocated to each of 12 treatment combinations. Blood pressure measurements were made on each subject after treatment, leading to the data below.

Biofeedback	Drug	Special diet No	Yes
Present	X	170 175 165 180 160 158	161 173 157 152 181 190
	Y	186 194 201 215 219 209	164 166 159 182 187 174
	Z	180 187 199 170 204 194	162 184 183 156 180 173
Absent	X	173 194 197 190 176 198	164 190 169 164 176 175
	Y	189 194 217 206 199 195	171 173 196 199 180 203
	Z	202 228 190 206 224 204	205 199 170 160 179 179

1. Find the analysis of variance table for the data and interpret the results.

2. Construct appropriate graphics as an aid to this interpretation.

3. Re-analyze the data after taking a log transformation and compare the results with those in step 1.

Exercise 5.4 The data in the **genotypes** data set are from a foster feeding experiment with rat mothers and litters of four different genotypes: A, B, I, and J. The measurement is the litter weight (in grams) after a trial feeding period. Investigate the effect of genotype of mother and litter on litter weight.

Litter genotype	Mother genotype	weight
A	A	61.5
A	A	68.2
A	A	64.0
A	A	65.0
A	A	59.7
A	B	55.0
A	B	42.0
A	B	60.2
A	I	52.5
A	I	61.8
A	I	49.5
A	I	52.7
A	J	42.0
A	J	54.0
A	J	61.0
A	J	48.2
B	J	39.6
B	A	60.3
B	A	51.7
B	A	49.3
B	A	48.0
B	B	50.8
B	B	64.7
B	B	61.7
B	B	64.0
B	B	62.0
B	I	56.5
B	I	59.0

(continued)

Exercise 5.4 (*continued*)

Litter genotype	Mother genotype	weight
B	I	47.2
B	I	53.0
B	J	51.3
B	J	40.5
I	A	37.0
I	A	36.3
I	A	68.0
I	B	56.3
I	B	69.8
I	B	67.0
I	I	39.7
I	I	46.0
I	I	61.3
I	I	55.3
I	I	55.7
I	J	50.0
I	J	43.8
I	J	54.5
J	A	59.0
J	A	57.4
J	A	54.0
J	A	47.0
J	B	59.5
J	B	52.8
J	B	56.0
J	I	45.2

(*continued*)

Exercise 5.4 (*continued*)

Litter genotype	Mother genotype	weight
J	I	57.0
J	I	61.4
J	J	44.8
J	J	51.5
J	J	53.0
J	J	42.0
J	J	54.0

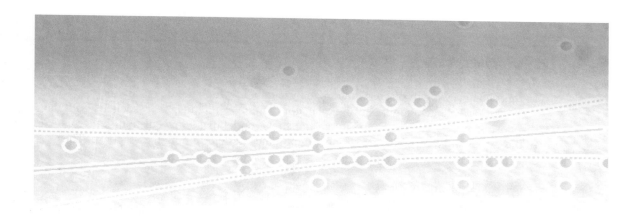

Chapter 6

Multiple Linear Regression

6.1 Introduction

In this chapter, we will discuss how to analyze data in which there is a continuous response variable and a number of explanatory variables that may be associated with the response variable. The aim is to build a statistical model that allows us to discover which of the explanatory variables are of most importance in determining the response. The statistical topics covered are:

- Multiple regression
- Interpretation of regression coefficients
- Regression diagnostics

6.2 Example: Consuming Ice Cream

The data shown in Table 6.1 were collected in a study to investigate how price and temperature influence consumption of ice cream. Here, we have a *response variable*: consumption of ice cream, and two *explanatory variables* (often inappropriately labeled *independent variables*): price and temperature. The aim is to fit a suitable statistical model to the data that allows us to determine how consumption of ice cream is affected by the other two variables.

Table 6.1 Ice Cream Consumption: Measured Over 30 4-Week Periods

Observation	Consumption	Price	Mean Temperature
1	.386	.270	41
2	.374	.282	56
3	.393	.277	63
4	.425	.280	68
5	.406	.272	69
6	.344	.262	65
7	.327	.275	61
8	.288	.267	47

(*continued*)

Table 6.1 (*continued*)

Observation	Consumption	Price	Mean Temperature
9	.269	.265	32
10	.256	.277	24
11	.286	.282	28
12	.298	.270	26
13	.329	.272	32
14	.318	.287	40
15	.381	.277	55
16	.381	.287	63
17	.470	.280	72
18	.443	.277	72
19	.386	.277	67
20	.342	.277	60
21	.319	.292	44
22	.307	.287	40
23	.284	.277	32
24	.326	.285	27
25	.309	.282	28
26	.359	.265	33
27	.376	.265	41
28	.416	.265	52
29	.437	.268	64
30	.548	.260	71

6.2.1 The Ice Cream Data: An Initial Analysis Using Scatterplots

The ice cream data in Table 6.1 are in a comma-separated file, **icecream.csv**, with the names of the variables, also comma-separated, in its first line. The data need to be imported to a SAS data set before they can be analyzed, but files of this type are very straightforward:

1. Select **File≻Import Data≻Local Computer**, browse to the location of the file **c:\saseg\data**, and click **Open**.

2. Under **Region to import**, check the box labeled **Specify line to use as column headings**. Line 1 is the default for this. The default text format is comma-delimited.

3. Click **Run.**

Some scatterplots of the three variables will be helpful in an initial examination of the data. For the scatterplots:

1. Select **Graph≻Scatter Plot**.

2. Under **Task Roles**, assign **consumption** to **Vertical** and **price** to **Horizontal**.

3. Click **Run.**

Repeat this with **temperature** having the role of **Horizontal** variable.

The resulting scatterplots are shown in Figures 6.1 and 6.2. The plots suggest that temperature is more influential than price in determining ice cream consumption with consumption increasing markedly as temperature increases.

Figure 6.1 Scatterplot of Ice Cream Consumption against Price

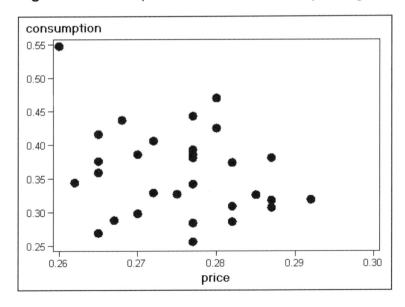

Figure 6.2 Scatterplot of Ice Cream Consumption against Temperature

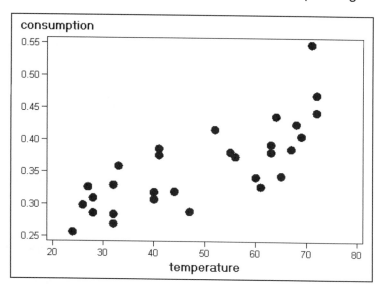

6.2.2 Ice Cream Sales: Are They Most Affected by Price or Temperature? How to Tell Using Multiple Regression

In Chapter 4, we examined the simple linear regression model that allows the effect of a single explanatory variable on a response variable to be assessed. We now need to extend the simple linear regression model to situations where there is more than a single explanatory variable. For continuous response variables, a suitable model is often *multiple linear regression*, which mathematically can be written as:

$$y_i = \beta_0 + \beta_1 x_{i1} + \beta_2 x_{i2} + \ldots\ldots + \beta_p x_{ip} + \varepsilon_i$$

where the $y_i, i = 1, n$ are the observed values of the response variable, and $x_{1i}, x_{2i}, \ldots, x_{pi}$, $i = 1, n$ are the observed values of the p explanatory variables; n is the number of observations in the sample, and the ε_i are the "error" terms in the model. The *regression coefficients*, $\beta_1, \beta_2, \ldots, \beta_p$, give the amount of change in the response variable associated with a unit change in the corresponding explanatory variable, *conditional*, on the other explanatory variables remaining unchanged. The regression coefficients are estimated from the sample data by *least squares*. For full details see, for example, Everitt (1996). The error terms in the model are assumed to have a normal distribution with mean zero and variance σ^2. The assumed normal distribution for the error terms in the model implies that, for given values of the explanatory variables, the response variable is normally distributed with a mean that is a linear function of the explanatory variables and a variance that is not dependent on the explanatory variables. The variation in the response variable can be partitioned into a part due to regression on the explanatory variables and a residual. The various terms in the partition of the variation of the response variable can be arranged in an analysis of variance table and the F-test of the equality of the regression variance (or *mean square*) and the residual variance gives a test of the hypothesis that there is no regression on the explanatory variables; i.e., the hypothesis that all the population regression coefficients are zero. Further details are available in Everitt (1996).

We now fit the multiple regression model to the ice cream data. Here, there are two explanatory variables: temperature and price. The multiple regression model is fitted to the data:

1. Select **Analyze≻Regression≻Linear**.

2. Under **Task Roles**, assign **consumption** the role of **Dependent variable** and **price** and **temperature** the role of **Explanatory variables** (Display 6.1). Note that there is no role of **Classification variable**. The **Explanatory variables** are all assumed to be quantitative.

3. Click **Run.**

Display 6.1 Task Roles Pane for Linear Regression of Ice Cream Data

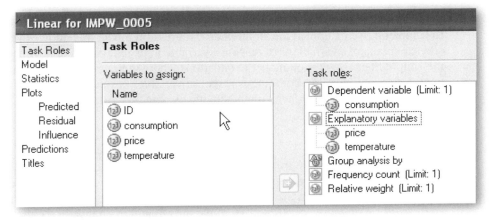

The results are shown in Table 6.2. The F-test in the analysis of variance table takes the value 23.27 and has an associated *p*-value that is very small. Clearly, the hypothesis that both regression coefficients are zero is not tenable. The multiple correlation coefficient gives the correlation between the observed values of the response variable (ice cream consumption) and the values predicted by the fitted model; the square of the coefficient (0.63) gives the proportion of the variance in ice cream consumption accounted for by price and temperature. The negative regression coefficient for price indicates that, for a given mean temperature, consumption decreases with increasing price. But as indicated by the two scatterplots in Section 6.3.1, only the estimated regression coefficient associated with temperature is statistically significant. The regression coefficient of consumption on temperature is estimated to be 0.00303 with an estimated standard error of 0.00047. A one-degree rise in temperature is estimated to increase consumption by 0.00303 units, conditional on price.

Table 6.2 Results from Applying the Multiple Regression Model to the Ice Cream Consumption Data

Number of Observations Read	30
Number of Observations Used	30

Analysis of Variance					
Source	**DF**	**Sum of Squares**	**Mean Square**	**F Value**	**Pr > F**
Model	2	0.07943	0.03972	23.27	<.0001
Error	27	0.04609	0.00171		
Corrected Total	29	0.12552			

Root MSE	0.04132	**R-Square**	0.6328
Dependent Mean	0.35943	**Adj R-Sq**	0.6056
Coeff Var	11.49484		

Parameter Estimates						
Variable	Label	DF	Parameter Estimate	Standard Error	t Value	Pr > |t|
Intercept	Intercept	1	0.59655	0.25831	2.31	0.0288
Price	Price	1	-1.40176	0.92509	-1.52	0.1413
temperature	temperature	1	0.00303	0.00046995	6.45	<.0001

6.2.3 Diagnosing the Multiple Regression Model Fitted to the Ice Cream Consumption Data: The Use of Residuals

An important final stage when fitting a multiple regression model is to investigate whether the assumptions made by the model, such as constant variance and normality of error terms, are reasonable. One way to assess the normality and constant variance assumptions is to look at the *residuals* from the model-fitting process. In their basic form, residuals are defined as:

Residual=observed response value-fitted response value

Various ways of plotting the residuals can be helpful in assessing particular components of the regression model. The most useful plots are as follows:

- A histogram or stem-and-leaf plot of the residuals can be useful checking for symmetry and specifically for the normality of the error terms in the regression model.

- Plot the residuals against corresponding values of each explanatory variable. Any sign of a nonlinear relationship in any plot might suggest that a higher order term (e.g., a quadratic) might be necessary for the particular explanatory variable.

- Plot the residuals against the *fitted* values of the response variable (i.e., the values predicted from the model). If the variance of the residuals appears to increase with the fitted values, a *transformation* of the response may be necessary before refitting the model.

Figure 6.3 shows some idealized residual plots that illustrate each of bullet points above.

Figure 6.3 Idealized Residual Plots

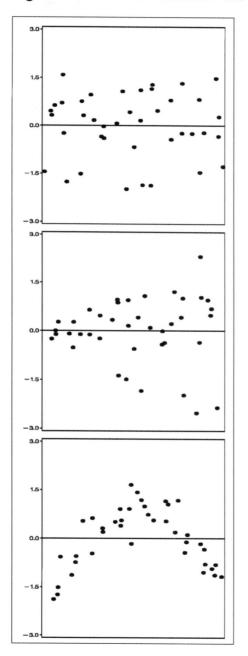

The simple residuals defined above can be shown to have unequal variances and to be slightly correlated. This correlation makes them less than ideal for detecting problems with fitted models, and the basic residuals are often *standardized* in some way before being used in a graphical examination of the regression model. Details are given in Rawlings et al. (2001).

To produce plots of residuals from the ice cream data:

1. Reopen the **Linear Regression** task (double-click or right-click **Open**).

2. Under **Plots≻Residual**, check **Standardized vs predicted Y** and **Standardized vs independents** (Display 6.2).

3. Under **Predictions**, check **Original sample** in **Data to predict** and **Residuals** in **Additional statistics** (Display 6.3).

4. Click **Run.**

5. Replace the results of the previous run.

Display 6.2 Selection of Residual Plots for Ice Cream Data

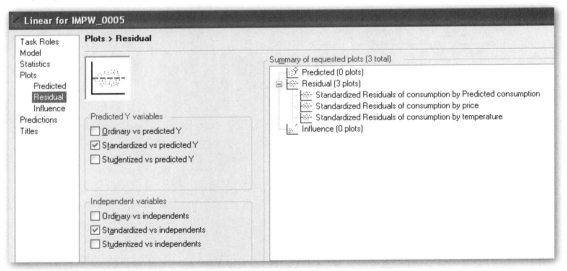

Display 6.3 Saving Residuals from Regression of the Ice Cream Data

As well as the residual plots shown in Figure 6.4, a data set is created which contains the predicted values and residuals and which can be used to check the normality of the residuals using the Distribution Analysis task.

1. Select the resulting data set, labeled **Linear regression…**.

2. Select **Describe≻Distribution Analysis**.

3. The **Task Roles** pane (Display 6.4) lists several variables in addition to the original variables: one for the values of **consumption** predicted by the model and three for different types of residuals. **Residual_consumption** is the raw residual, i.e., the observed value minus the predicted value; **student_consumption** is the standardized residual, i.e., the raw residual scaled to have a mean of zero and standard deviation of one. The third, **rstudent_consumption**, is similar to the standardized residual but with the current observation omitted from the calculations. Assign **student_consumption** as the **Analysis variable**.

4. Under **Distributions≻Normal**, check **Normal** to test for normality.

5. Under **Plots**, select **Histogram Plot**.

6. Click **Run.**

Display 6.4 Task Roles Pane Showing Additional Variables in the Prediction Data Set

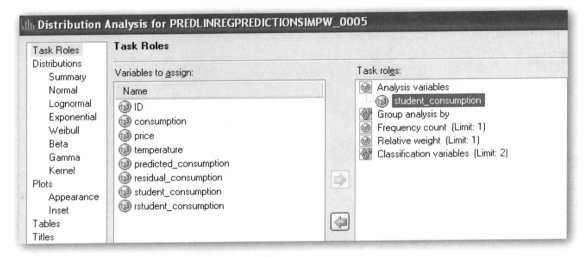

The results are given in Figures 6.4 and 6.5. On the whole, the plots all appear satisfactory and give little cause for concern about the assumptions made in fitting the model or for the results obtained for the fitting procedure.

Figure 6.4 Standardized Residuals for the Ice Cream Consumption Data Plotted against Price and Temperature

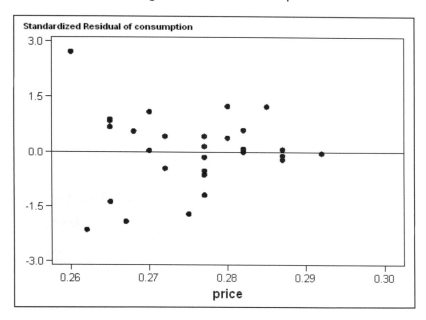

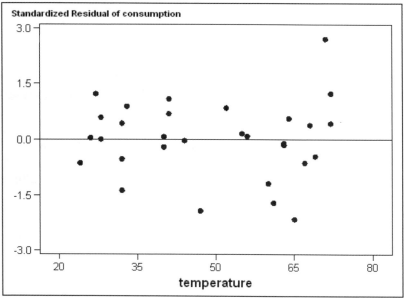

Figure 6.5 Histogram of Standardized Residuals for Ice Cream Consumption Data

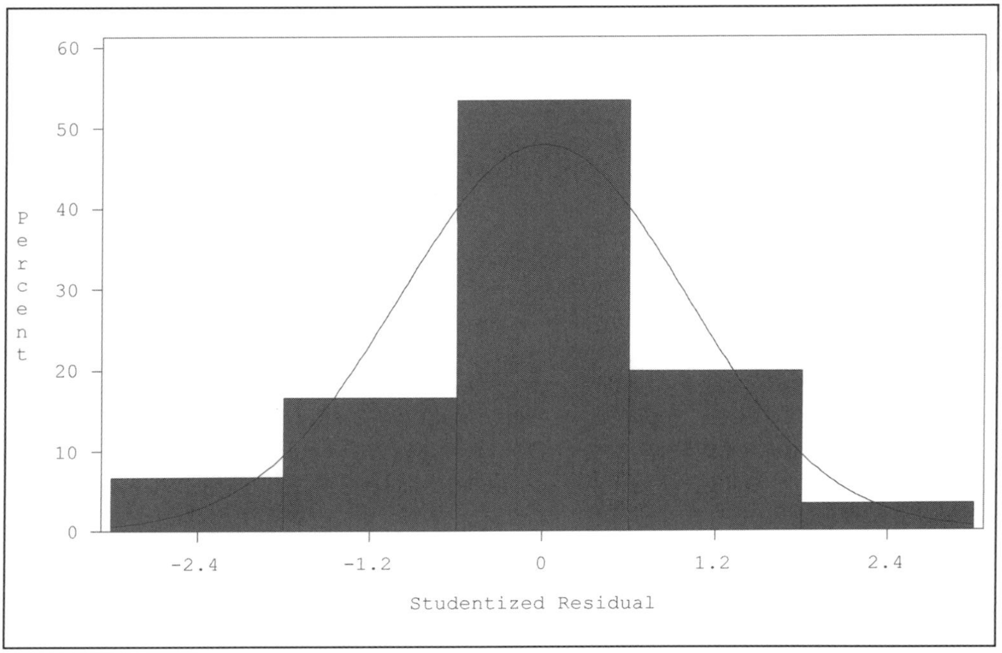

6.3 Example: Making It Rain by Cloud Seeding

Weather modification, or cloud seeding, is the treatment of individual clouds or storm systems with various inorganic or organic materials in the hope of achieving an increase in rainfall. Introduction of such material into a cloud that contains supercooled water, that is liquid water colder than zero degrees Celsius, has the aim of inducing freezing, with the consequent ice particles growing at the expense of liquid droplets and becoming heavy enough to fall as rain from the clouds that otherwise would produce none.

The data in Table 6.3 were collected in the summer of 1975 from an experiment to investigate the use of massive amounts of silver iodine (100 to 1000 grams per cloud) in cloud seeding to increase rainfall (Woodley et al. 1977). In the experiment, which was conducted in an area of Florida, 24 days were judged suitable for seeding on the basis that a measured suitability criterion was met. The suitability criterion (S-NE), which is defined in detail in Woodley et al. biases the decision for experimentation against naturally rainy days. On thus defined suitable days, a decision was taken at random as to whether to seed or not. The aim in analyzing the cloud seeding data is to see how rainfall is related to the other variables and, in particular, to determine the effectiveness of seeding.

Table 6.3 Cloud Seeding Data

Seeding	Time	S-NE	Cloudcover	Pre-wetness	Echo motion	Rainfall
0	0	1.75	13.40	0.274	1	12.85
1	1	2.70	37.90	1.267	0	5.52
1	3	4.10	3.90	0.198	1	6.29
0	4	2.35	5.30	0.526	0	6.11
1	6	4.25	7.10	0.250	0	2.45
0	9	1.60	6.90	0.018	1	3.61
0	18	1.30	4.60	0.307	0	0.47
0	25	3.35	4.90	0.194	0	4.56
0	27	2.85	12.10	0.751	0	6.35
1	28	2.20	5.20	0.084	0	5.06
1	29	4.40	4.10	0.236	0	2.76
1	32	3.10	2.80	0.214	0	4.05
0	33	3.95	6.80	0.796	0	5.74
1	35	2.90	3.00	0.124	0	4.84
1	38	2.05	7.00	0.144	0	11.86
0	39	4.00	11.30	0.398	0	4.45
0	53	3.35	4.20	0.237	1	3.66
1	55	3.70	3.30	0.960	0	4.22
0	56	3.80	2.20	0.230	0	1.16
1	59	3.40	6.50	0.142	1	5.45
1	65	3.15	3.10	0.073	0	2.02
0	68	3.15	2.60	0.136	0	0.82
1	82	4.01	8.30	0.123	0	1.09
0	83	4.65	7.40	0.168	0	0.28

6.3.1 The Cloud Seeding Data: Initial Examination of the Data Using Box Plots and Scatterplots

For the cloud seeding data, we will construct box plots of the rainfall in each category of the dichotomous explanatory variables (**seeding** and **echomotion**) and scatterplots of rainfall against each of the continuous explanatory variables (**cloudcover**, **sne**, **prewetness**, and **time**). We first create a new process flow for the analysis of these data.

1. Select **File≻New≻Process Flow**.

2. Rename the process flow **cloud seeding** (right-click on the **Process Flow** tab and select **Rename**).

The data are stored in a tab-separated file, **cloud.tab**, with the variable names in the first line. To import the data:

1. Select **File≻Import Data≻Local Computer**, browse to the location **c:\saseg\data**, and select and open **cloud.tab**.

2. Under **Region to import**, check **Specify line to use as column headings**. Line 1 is the default.

3. Under **Text Format**, select **Delimited** and **Tab**.

4. Click **Run**.

For the box plots:

1. Select **Graph≻Box Plot**.

2. Assign **rainfall** to **Vertical** and **seeding** to **Horizontal**.

3. Click **Run**.

Repeat with **echomotion** as the horizontal variable.

The plots are shown in Figures 6.6 through 6.11. Both the box plots and the scatterplots show some evidence of two outliers. In particular, the scatterplot of rainfall against cloud cover suggests one very clear outlying observation which on inspection turns out to be the second observation in the data set. For the time being, we shall not remove any observations but simply bear in mind during the modeling process to be described later that outliers may cause difficulties.

Figure 6.6 Box Plots of Rainfall for Seeding and Not Seeding Days

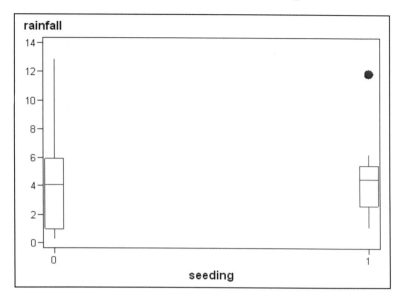

Figure 6.7 Box Plots of Rainfall for Moving and Stationary Echomotion

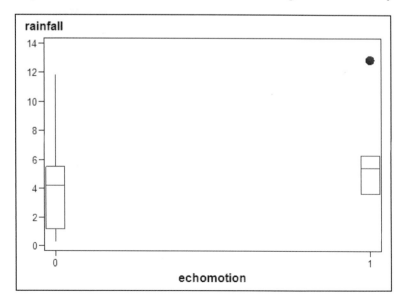

Figure 6.8 Scatterplot of Rainfall against Time

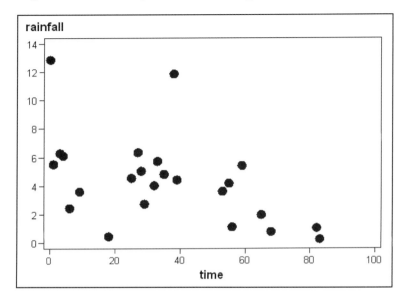

Figure 6.9 Scatterplot of Rainfall against S-NE Criterion

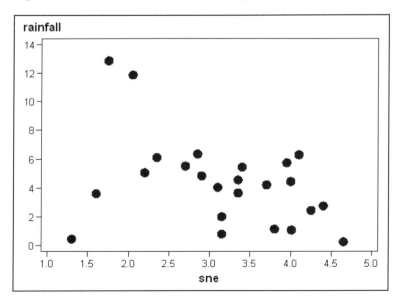

Figure 6.10 Scatterplot of Rainfall against Cloud Cover

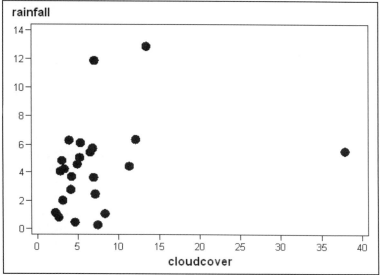

Figure 6.11 Scatterplot of Rainfall against Prewetness

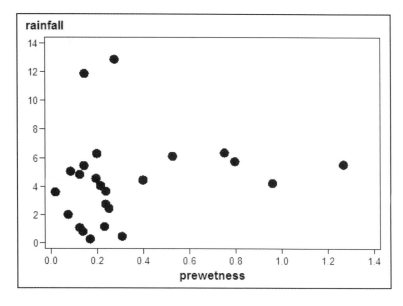

6.3.2 When Is Cloud Seeding Best Carried Out? How to Tell Using Multiple Regression Models Containing Interaction Terms

For the cloud seeding data, one thing to note about the explanatory variables is that two of them, seeding and echomotion, are binary variables. Should such an explanatory variable be allowed in a multiple regression model? In fact, there is really no problem in including such variables in the model since, in a strict sense, the explanatory variables are assumed to be fixed rather than random. In practice, of course, fixed explanatory variables are rarely the case and so the results from a multiple regression analysis are interpreted as being conditional on the observed values of the explanatory variables; this rather arcane point is discussed in more detail in Everitt (1996). It is only the response variable that is considered to be random.

In the cloud seeding example, there are theoretical reasons to consider a particular model for the data (Woodley et al. 1977), namely one in which the effect of some of the other explanatory variables is modified by seeding. So, the model we will consider is one that allows *interaction terms* for seeding with each of the other explanatory variables except time.

For the analysis of the ice cream data in the previous section, we used the linear regression task (**Analyze≻Regression≻Linear**) to do the analysis. It is also possible to do multiple regression using the Linear Models task (**Analyze≻ANOVA≻Linear Models**) introduced in Chapter 5. The choice between them is largely a matter of which is more convenient for the particular analysis as they offer somewhat different features. The Linear Models task can accommodate both categorical (with more than two categories) and continuous predictors and interactions can be specified in the model.

The Linear Regression task can accommodate only continuous predictors and binary variables. For categorical variables with more than two categories, *dummy variables* would need to be derived beforehand (Everitt, 1996, and Chapter 7 of this book). To enter an interaction term between two variables in Linear Regression, a new variable needs to be calculated as the product of the two variables, and this new variable needs to be entered into the model. An advantage of the Linear Regression task is that it offers a number of *variable selection methods*, whereas Linear Models does not. For details of variable selection methods see Everitt (1996).

Since the model to be fitted to the cloud seeding data requires a number of interactions, the Linear Models task will be more convenient and this will therefore be used.

1. Select **Analyze≻ANOVA≻Linear Models**.

2. Under **Task Roles**, **rainfall** is the **Dependent variable**, and the remaining variables are treated as **Quantitative variables**.

3. Under **Model**, select all the variables, and click **Main** to enter their main effects.

4. To enter the required interactions with **seeding**, first click on **seeding**, then CTRL-click on the other variable in the interaction, and then click the **Cross** button. Repeat this for each interaction with **seeding**. When all the terms have been entered, the **Model** panel should look like Display 6.5.

5. Click **Run**.

Display 6.5 Model Panel for Cloud Seeding Data

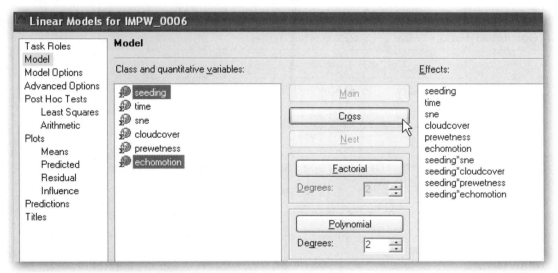

The results of fitting the specified model are shown in Table 6.4. The first thing to note about Table 6.4 is the presence of both Type I and Type III sums of squares. As we would expect from our discussion of both types of sums of squares in the previous chapter, they lead to different p-values for assessing the statistical significance of the estimated regression coefficients. We should also note that the p-values for the Type III sums of squares are identical to the p-values given for each estimated regression coefficient found from a t-test of the hypothesis that the corresponding population coefficient is zero (the t-statistic is simply the regression coefficient estimate divided by its estimated standard error). The p-values are identical because the regression coefficients are estimated conditional on all other terms in the model, and the Type III sums of squares are essentially found in the same way. Here, where we are primarily interested in the interaction effects of other explanatory variables with seeding, it is appropriate to use the Type III sums of squares or equivalently the t-tests on each coefficient to judge which interactions are of most importance. But if we wanted to assess the main effects in the model, then Type III sums of squares would not be the ones to use as argued in Chapter 5.

Table 6.4 Results of Fitting a Multiple Regression Model with Interactions to the Cloud Seeding Data

Source	DF	Sum of Squares	Mean Square	F Value	Pr > F
Model	10	159.1460026	15.9146003	3.27	0.0243
Error	13	63.1888933	4.8606841		
Corrected Total	23	222.3348958			

R-Square	Coeff Var	Root MSE	rainfall Mean
0.715794	50.07353	2.204696	4.402917

Source	DF	Type I SS	Mean Square	F Value	Pr > F
Seeding	1	1.28343750	1.28343750	0.26	0.6160
Time	1	55.31045354	55.31045354	11.38	0.0050
Sne	1	11.49594672	11.49594672	2.37	0.1481
Cloudcover	1	2.00314031	2.00314031	0.41	0.5321
Prewetness	1	0.33184258	0.33184258	0.07	0.7980
Echomotion	1	15.15890800	15.15890800	3.12	0.1009
seeding*sne	1	33.15823806	33.15823806	6.82	0.0215
seeding*cloudcover	1	38.82083001	38.82083001	7.99	0.0143
seeding*prewetness	1	1.36347743	1.36347743	0.28	0.6053
seeding*echomotion	1	0.21972842	0.21972842	0.05	0.8349

Source	DF	Type III SS	Mean Square	F Value	Pr > F
Seeding	1	60.47295296	60.47295296	12.44	0.0037
Time	1	15.66429701	15.66429701	3.22	0.0959
Sne	1	1.20110501	1.20110501	0.25	0.6274
Cloudcover	1	15.40695933	15.40695933	3.17	0.0984
Prewetness	1	6.32677854	6.32677854	1.30	0.2745
Echomotion	1	12.93729038	12.93729038	2.66	0.1268
seeding*sne	1	30.94793564	30.94793564	6.37	0.0254
seeding*cloudcover	1	19.77764964	19.77764964	4.07	0.0648
seeding*prewetness	1	1.58289424	1.58289424	0.33	0.5780
seeding*echomotion	1	0.21972842	0.21972842	0.05	0.8349

Parameter	Estimate	Standard Error	t Value	Pr > |t|
Intercept	-0.34624093	2.78773403	-0.12	0.9031
Seeding	15.68293481	4.44626606	3.53	0.0037
Time	-0.04497427	0.02505286	-1.80	0.0959
Sne	0.41981393	0.84452994	0.50	0.6274
Cloudcover	0.38786207	0.21785501	1.78	0.0984
Prewetness	4.10834188	3.60100694	1.14	0.2745
Echomotion	3.15281358	1.93252592	1.63	0.1268
seeding*sne	-3.19719006	1.26707204	-2.52	0.0254
seeding*cloudcover	-0.48625492	0.24106012	-2.02	0.0648
seeding*prewetness	-2.55706696	4.48089584	-0.57	0.5780
seeding*echomotion	-0.56221845	2.64429975	-0.21	0.8349

The tests of the interactions in the model suggest that the interaction of seeding with the S-NE criterion significantly affects rainfall. A suitable graph will help in the interpretation of the significant seeding x S-NE criterion interaction. For this graph, we can use the Line Plot task to show the relationship between rainfall and suitability for seeding (S-NE) on days on which seeding did and did not occur.

1. Select **Graph➤Line Plot**.

2. Under **Line Plot**, select **Multiple line plots by group column** as the type of plot.

3. Under **Task Roles**, assign **rainfall** to **Vertical**, **sne** to **Horizontal**, and **seeding** to **Group charts by**.

4. Under **Appearance➤Plots**, decide how the data for seeding and non-seeding days are distinguished in the plot. The default is to use different colors with the same line type and plotting symbol. Here, we will illustrate the use of different line types and plotting symbols. The panel in the center of the pane shows the two values of seeding, 0 and 1. Highlighting either of these (by clicking on it) shows how the line and data points will be plotted for that group of observations. A solid line is the default and a circle as the plotting symbol. Leaving the values for seeding=0 at their defaults, we change the values for seeding=1 to a dashed line and triangle as the plotting symbol (see Display 6.6).

5. Under **Interpolations**, select **regression** for both subgroups (seeding: 0 and 1) which will fit and draw a separate regression line for each group.

6. Click **Run.**

The result is shown in Figure 6.12.

Display 6.6 Setting Plot Options for Days on Which Seeding Took Place

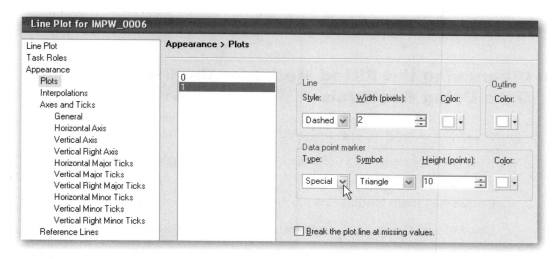

Figure 6.12 Scatterplot of Rainfall versus S-NE for Seeding and Non-Seeding Days

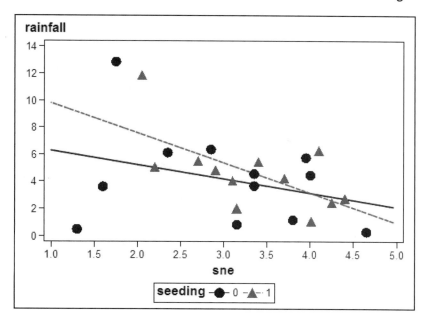

The plot suggests that for smaller S-NE values, seeding produces greater rainfall than no seeding, whereas seeding tends to produce less rainfall for larger S-NE values. The crossover point is at an S-NE value of approximately 4 which might suggest that, for most success, seeding should be applied when the S-NE criterion is less than 4.

6.3.3 Diagnosing the Fitted Model for the Cloud Seeding Data Using Residuals

For the cloud seeding data, we will plot residuals against predicted value and produce a probability plot of residuals to examine their normality.

The process is very similar to that for the ice cream data, even though a different task has been used for the analysis.

1. Reopen the **Linear Models** task (double-click or right-click **Open**).

2. Under **Plots≻Residual**, check **Standardized vs. predicted Y**.

3. Under **Predictions**, check **Original data** as the **Data to predict** and **Residuals** as **Additional statistics**.

4. Click **Run**.

5. Replace the results of the previous run.

A new data set is created containing the predicted values and residuals as well as the original variables, and this new data set is used to check the normality of the residuals.

1. Make the new data set active (by clicking on it).

2. Select **Describe≻Distribution Analysis**.

3. Under **Task Roles, student_rainfall** is the **Analysis variable**.

4. Under **Distributions≻Normal**, check **Normal** (we also changed the line color to black here).

5. Under **Plots**, select **Probability Plot**.

6. Click **Run**.

The results are shown in Figures 6.13 and 6.14. The plot of residuals against fitted values suggest that two observations (numbers 1 and 15) are outliers, and it may be of interest to refit the model with the outliers removed (see Exercise 6.2). The normal probability plot shows little evidence of any worrying departure from normality.

Figure 6.13 Plot of Standardized Residuals versus Fitted Rainfall for the Model Fitted to the Cloud Seeding Data

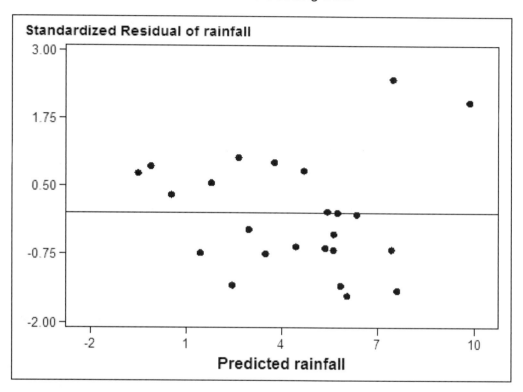

Figure 6.14 Probability Plot of Standardized Residuals from the Model Fitting of the Cloud Seeding Data

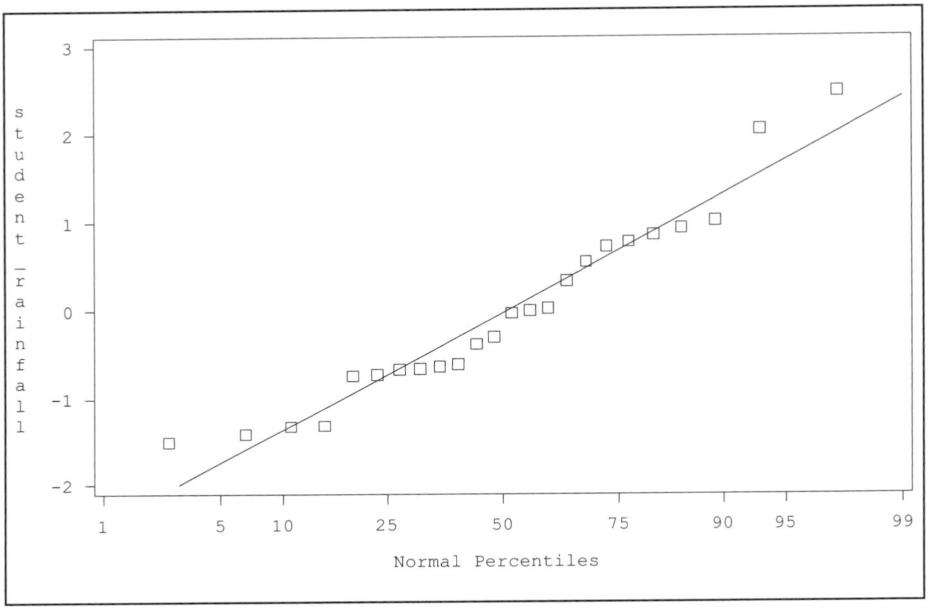

6.4 Exercises

Exercise 6.1 For the ice cream consumption data, investigate what happens when you fit a simple linear regression of consumption on price and then add temperature to the model. Repeat this exercise fitting first temperature and then adding price.

Exercise 6.2 Repeat the analysis of the cloud seeding data after removing observations 1 and 15. Compare your results with those given in the text.

Exercise 6.3 The data in the **fat** data set are taken from a study investigating a new method of measuring body composition and giving the body fat percentage, age, and sex for 20 normal adults aged between 23 and 61 years.

1. Construct a scatterplot percentage fat against age labeling the points according to sex.

2. Fit a multiple regression model to the data using **Fat** as the response variable and **Age** and **Sex** as explanatory variables. Interpret the results with the help of a scatterplot showing the essential features of the fitted model.

3. Fit a further model which allows an interaction between **Age** and **Sex** and again construct a diagram that will help you interpret the results.

Age	Sex	% Fat
23	M	9.5
23	F	27.9
27	M	7.8
27	M	17.8
39	F	31.4
41	F	25.9
45	M	27.4
49	F	25.2
50	F	31.1
53	F	34.7
53	F	42.0
54	F	29.1
56	F	32.5
57	F	30.3
58	F	33.0
58	F	33.8
60	F	41.1
61	F	34.5

Exercise 6.4 The data in the Microsoft Office Excel spreadsheet usair.xls, mentioned in Chapter 1, relate to air pollution in 41 U.S. cities. Seven variables are recorded for each of the cities:

- SO_2 content of air in micrograms per cubic meter

- Average annual temperature in °F

- Number of manufacturing enterprises employing 20 or more workers

- Population size (1970 census) in thousands

- Average annual wind speed in miles per hour

- Average annual precipitation in inches

- Average number of days with precipitation per year

so_2	temperature	factories	population	wind	rain	rainydays
10	70.3	213	582	6	7.05	36
13	61	91	132	8.2	48.52	100
12	56.7	453	716	8.7	20.66	67
17	51.9	454	515	9	12.95	86
56	49.1	412	158	9	43.37	127
36	54	80	80	9	40.25	114
29	57.3	434	757	9.3	38.89	111
14	68.4	136	529	8.8	54.47	116
10	75.5	207	335	9	59.8	128
24	61.5	368	497	9.1	48.34	115
110	50.6	3344	3369	10.4	34.44	122
28	52.3	361	746	9.7	38.74	121
17	49	104	201	11.2	30.85	103
8	56.6	125	277	12.7	30.58	82
30	55.6	291	593	8.3	43.11	123
9	68.3	204	361	8.4	56.77	113
47	55	625	905	9.6	41.31	111
35	49.9	1064	1513	10.1	30.96	129
29	43.5	699	744	10.6	25.94	137
14	54.5	381	507	10	37	99
56	55.9	775	622	9.5	35.89	105
14	51.5	181	347	10.9	30.18	98
11	56.8	46	244	8.9	7.77	58
46	47.6	44	116	8.8	33.36	135
11	47.1	391	463	12.4	36.11	166
23	54	462	453	7.1	39.04	132

(*continued*)

SO$_2$	temperature	factories	population	wind	rain	rainydays
65	49.7	1007	751	10.9	34.99	155
26	51.5	266	540	8.6	37.01	134
69	54.6	1692	1950	9.6	39.93	115
61	50.4	347	520	9.4	36.22	147
94	50	343	179	10.6	42.75	125
10	61.6	337	624	9.2	49.1	105
18	59.4	275	448	7.9	46	119
9	66.2	641	844	10.9	35.94	78
10	68.9	721	1233	10.8	48.19	103
28	51	137	176	8.7	15.17	89
31	59.3	96	308	10.6	44.68	116
26	57.8	197	299	7.6	42.59	115
29	51.1	379	531	9.4	38.79	164
31	55.2	35	71	6.5	40.75	148
16	45.7	569	717	11.8	29.07	123

Use multiple regression to investigate which of the other variables most determine pollution as indicated by the SO$_2$ content of the air. (Preliminary investigation of the data may be necessary to identify possible outliers and pairs of explanatory variables that are so highly correlated that they may cause problems for model fitting.)

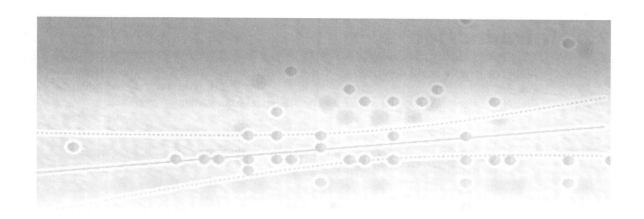

Chapter 7

Logistic Regression

7.1 Introduction

In this chapter, we will describe how to deal with data where there are a binary response variable and a number of explanatory variables. The aim is to see how the explanatory variables affect the response variable. The statistical topics to be covered are:

- Regression model for a binary response variable: Logistic regression
- What the logistic regression model tells us: Interpretation of regression coefficients and odds ratios

7.2 Example: Myocardial Infarctions

The data in Table 7.1 come from a study described in Kasser and Bruce (1969). A total of 117 male coronary patients were studied to try to determine how history of past myocardial infarctions is dependent on the other variables observed.

Table 7.1 Data on History of Past Myocardial Infarctions

42 2 1 1 0	50 0 0 1 0	35 2 1 1 0	55 3 1 1 1
66 2 1 1 0	72 3 0 1 1	34 2 1 1 0	51 1 0 1 0
56 2 1 1 0	56 3 1 1 0	68 3 1 1 0	46 1 1 1 0
55 2 1 1 0	56 3 1 1 0	49 3 0 1 0	69 1 0 1 0
41 2 1 1 0	63 2 1 1 0	55 2 0 1 1	51 3 1 1 1
62 0 1 0 1	53 1 1 1 0	58 0 1 1 0	49 1 0 1 1
46 2 1 1 1	53 0 1 0 0	43 2 1 1 0	58 3 1 1 0
44 2 0 1 1	57 3 0 1 0	39 2 0 1 0	38 3 1 1 0
50 1 0 1 0	57 1 0 1 1	66 3 0 1 1	50 1 1 0 0
73 3 0 1 0	62 2 1 1 0	50 2 1 1 0	38 1 1 1 0
48 2 1 1 0	73 2 0 1 0	45 3 0 1 0	58 1 1 0 0
53 2 1 1 0	44 2 0 1 0	53 0 0 1 0	69 0 0 1 0
51 3 1 1 1	63 3 1 1 0	56 3 1 1 1	66 0 0 1 0
51 3 1 1 1	59 1 0 1 0	49 2 1 1 1	49 2 0 1 0

(continued)

Table 7.1 (*continued*)

59 0 0 1 1	51 1 1 0 0	49 0 1 0 0	62 0 1 1 0
54 3 1 1 1	52 3 1 0 0	56 2 1 1 0	44 0 0 1 1
41 2 1 1 1	64 0 1 0 0	38 0 1 1 0	58 3 1 1 0
56 2 1 0 1	53 2 1 0 0	39 0 1 1 0	45 2 1 1 0
38 0 0 1 1	58 1 0 1 0	62 2 1 1 0	58 3 1 1 0
40 3 1 1 0	53 0 1 1 1	70 3 1 1 1	54 2 1 1 0
42 1 1 1 0	58 2 0 1 0	53 2 1 1 1	55 2 1 1 1
51 1 0 1 0	45 1 1 1 1	68 2 1 1 0	68 2 1 1 0
52 1 1 1 0	42 3 0 1 0	50 2 0 0 1	68 2 1 1 0
37 0 0 1 0	60 2 0 1 1	46 2 0 1 0	47 1 1 1 1
48 1 1 0 0	34 1 1 0 1	58 3 1 1 0	55 0 0 1 0
35 0 1 1 0	64 2 1 1 0	57 2 0 0 0	
35 1 1 0 0	35 1 1 0 1	55 3 1 0 0	
48 3 0 1 1	42 2 0 1 1	52 0 1 1 0	
52 2 0 1 1	53 2 1 0 0	61 2 0 0 0	
46 2 0 1 1	58 1 1 0 1	45 2 0 0 0	
51 3 0 1 0	38 1 0 1 0	51 2 1 1 0	

Description of variables, in order:

1=age in years
2=function-functional class: none (0), minimal (1), moderate (2), more than moderate (3)
3=infarct-history of past myocardial infarctions: none (0). Present (1)
4=angina-history of angina pectoris: none (0) present (1)
5=high bp: history of high blood pressure: none (0) present (1)

7.2.1 Myocardial Infarctions: What Predicts a Past History of Myocardial Infarctions? Answering the Question Using Logistic Regression

The multiple regression model considered in the previous chapter is suitable for investigating how a continuous response variable depends on a set of explanatory variables. But, can it be adapted to model a *binary response variable*? For example, in Table 7.1, a patient's past history of myocardial infarctions is a binary response variable and we would like to investigate how it is affected by the other variables in the data set. A possible way to proceed is to consider modeling the probability that the binary response takes the value 1; that is, the probability that a patient has a history of myocardial infarctions in our particular example.

A little thought shows that the multiple regression model cannot help us here. Firstly, the assumption that the response is normally distributed conditional on the explanatory variables is clearly no longer justified. And there is another fundamental problem: the application of the multiple regression model to the probability that the binary response takes the value 1; this could lead to fitted values *outside* the range 0 to 1, and this is clearly unacceptable for the probability being modeled.

7.2.2 Odds

So with a binary response variable, we need to consider an alternative approach to multiple regression, and the most common alternative is known as *logistic regression*. Here, the logarithm of the *odds* of the response variable being one (often known as the *logit* transformation of the probability) is modeled as a linear function of the explanatory variables. What are odds? Simply, the ratio of the probability that the binary variable takes the value 1, to the probability that the variable takes the value 0. Representing the probability of a 1 as p so that the probability of a zero is $(1-p)$, then the odds is simply given by $p/(1-p)$. For example, when tossing an unbiased die, the odds of a six are 1/6 divided by 5/6, giving the value 1/5. An experienced gambler would say that the odds of a six are five to one against.

But back to the logistic regression model which in mathematical terms can be written as:

$$\log(\frac{p}{1-p}) = \beta_0 + \beta_1 x_1 + \beta_2 x_2 + ... \beta_p x_p$$

where $x_1, x_2, ..., x_p$ are the explanatory variables. Now as p varies between 0 and 1, the logit transformation of p varies between minus and plus infinity, thus removing directly one of the problems mentioned above. The model can be rewritten in terms of the probability p as:

$$p = \frac{\exp(\beta_0 + \beta_1 x_1 + \beta_2 x_2 + ... \beta_p x_p)}{1 + \exp(\beta_0 + \beta_1 x_1 + \beta_2 x_2 + ... \beta_p x_p)}$$

Full details of the distributional assumptions of the model and of how the parameters in the model are estimated are given in Der and Everitt (2005). Below however, we will concentrate on how to obtain estimates of the parameters, $\beta_0, \beta_1, \beta_2, ..., \beta_p$ using SAS Enterprise Guide and on how to interpret the estimates after we find them.

7.2.3 Applying the Logistic Regression Model with a Single Explanatory Variable

We shall apply the logistic regression model to the data in Table 7.1, first using only the single explanatory variable, **angina**; considering this simple model will serve to clarify some of the points raised above. The data are in a comma-separated file, **coronary.csv**, with the variable names in the first line. To read them into a SAS data set:

1. Select **File≻Import Data≻Local Computer**, navigate to **c:\saseg\data**, select it and click **Open**.

2. Under **Region to import**, check the box labeled **Specify line to use as column headings**. Line 1 is the default. Under **Text Format**, the default is **Delimited** and so it does not need to be changed.

3. Under **Column Options**, check that the variables have been correctly assigned.

4. Click **Run.**

For the logistic model:

1. Select **Analyze≻Regression≻Logistic**.

2. Under **Task Roles,** assign **mi** the role of **Dependent variable**. With **mi** highlighted, alter the **Response variable** sort order from **Ascending** to **Descending** by clicking on **Sort order** and use the drop-down menu (Display 7.1). By default, the lower value is treated as the response to be modeled; altering the sort order reverses this. Assign **angina** the role of **Quantitative variables.**

3. Under **Model**, select **angina**, and click **Main** to enter its main effect into the model.

4. Click **Run**.

The results are shown in Table 7.2.

Display 7.1 Task Roles Pane for Logistic Regression: Altering the Response Variable Sort Order

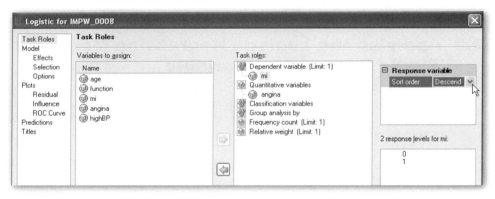

Table 7.2 Results of a Logistic Regression of the Coronary Data Using Angina as the Predictor

Model Information		
Data Set	WORK.SORTTEMPTABLESORTED	
Response Variable	mi	mi
Number of Response Levels	2	
Model	binary logit	
Optimization Technique	Fisher's scoring	

Number of Observations Read	117
Number of Observations Used	117

Logistic Regression Results

Response Profile		
Ordered Value	mi	Total Frequency
1	1	74
2	0	43

Probability modeled is mi=1.

Model Convergence Status
Convergence criterion (GCONV=1E-8) satisfied.

Model Fit Statistics		
Criterion	Intercept Only	Intercept and Covariates
AIC	155.884	154.140
SC	158.646	159.664
-2 Log L	153.884	150.140

Testing Global Null Hypothesis: BETA=0			
Test	Chi-Square	DF	Pr > ChiSq
Likelihood Ratio	3.7443	1	0.0530
Score	3.4512	1	0.0632
Wald	3.2368	1	0.0720

Analysis of Maximum Likelihood Estimates					
Parameter	DF	Estimate	Standard Error	Wald Chi-Square	Pr > ChiSq
Intercept	1	1.4469	0.5557	6.7791	0.0092
Angina	1	-1.0674	0.5933	3.2368	0.0720

Logistic Regression Results

Odds Ratio Estimates		
Effect	Point Estimate	95% Wald Confidence Limits
Angina	0.344	0.107 1.100

Association of Predicted Probabilities and Observed Responses			
Percent Concordant	20.8	Somers' D	0.137
Percent Discordant	7.2	Gamma	0.488
Percent Tied	72.0	Tau-a	0.064
Pairs	3182	C	0.568

7.2.4 Interpreting the Regression Coefficient in the Fitted Logistic Regression Model

For now, we will concentrate on how to interpret the estimated regression coefficient for angina (shown under **Analysis of Maximum Likelihood Estimates** in Table 7.2). The fitted model is:

log(odds of a past infarct)=1.447–1.067angina

So for patients with *no* past history of angina, that is angina=0, we have:

log(odds of a past infarct)=1.447

and for patients *with* a past history of angina, that is angina=1:

log(odds of a past infarct)=1.447–1.067

Taking the difference of the two log odds we have:

log(odds of an infarct for patients with angina)–log(odds of an infarct for patients without angina)=log(odds for patients with angina/odds for patients without angina)=-1.067

We can now see that for a dichotomous explanatory variable coded zero/one, the estimated regression coefficient is simply the log of the *odds ratio*. Rather than dealing with the log, we can find the odds ratio itself simply by exponentiating the regression coefficient. The odds ratio is conveniently provided in Table 7.2 in the **Odds Ratio Estimates** section of the table; in the Point Estimate column, the odds ratio is seen to take the value of 0.344.

How is the log odds value interpreted? The estimated odds ratio implies that the odds of patients *with* a past history of angina suffering an infarct are about a third that of patients *without* a previous history of angina. The finding appears to be somewhat curious but it may reflect that patients who suffer from angina are given some treatment which helps prevent an infarct. In addition, only 19 of the 117 patients have no previous history of angina. It must also be remembered that the 95% confidence interval for the odds ratio also given in Table 7.2, namely [0.107, 1.100], contains the value 1, a value that would correspond to the odds of an infarct for patients with and without angina being equal. (A value of 1 for the odds ratio is equivalent to a value 0 for the regression coefficient itself.) Consequently, there is little convincing evidence that a history of angina has any substantial effect on the occurrence of a previous infarct.

The **Testing Global Null Hypothesis** section of Table 7.2 provides three different statistics for assessing the hypothesis that the single regression coefficient in the model is 0; each statistic has an associated *p*-value greater than 0.05 confirming the inference made above from using the confidence interval, namely that there is no evidence that the regression coefficient differs from 0. Collett (2002) includes more details about the different model fit statistics.

7.2.5 Applying the Logistic Regression Model Using SAS Enterprise Guide

In a logistic regression model with more than a single explanatory variable, a regression coefficient associated with an explanatory variable gives the change in the log odds that the response variable takes the value 1 when the explanatory variable increases by one, conditional on the other explanatory variables remaining constant. So, exponentiating the coefficient gives the corresponding odds ratio for a change of one unit in the explanatory variable. For a continuous explanatory variable, the value of 1 will not be biologically very interesting. For example, an increase of 1 year in age or 1mm Hg in systolic blood pressure may be too small to be considered important. A change of 10 years or 10mm Hg might be considered more useful. (In some cases, a change of 1 is too large, and a change of 0.01 might be more realistic.) We show how to deal with the potential need to look at regression coefficients associated with changes of other than one unit in an explanatory variable below, after we have shown how to use SAS Enterprise Guide to apply the logistic regression model to the complete myocardial infarction data set considering all four explanatory variables:

1. Reopen the **Logistic** task (double-click or right-click **Open**).

2. Under **Task Roles**, assign **age** and **highbp** as **Quantitative variables**, but consider how to deal with the explanatory variable, **function**. This is essentially an ordered categorical variable with four categories. If you choose to use it in the modeling process as a quantitative variable with values 0, 1, 2, and 3, you are assuming that changes in the variable from, for example, 0 to 1 and from 2 to 3, have an equal effect on the response variable. This is unlikely to be the case. Consequently, you should choose to assign it the role as a **Classification variable**

and, with it highlighted, assign the coding style as **Reference**. The **Task Roles** pane should then look like Display 7.2.

3. Under **Model▸Effects**, select all variables and click **Main** to enter all their main effects into the model.

4. Click **Run.**

5. Do not replace the results of the previous run.

The results are shown in Table 7.3.

Display 7.2 Task Roles Pane for Coronary Data

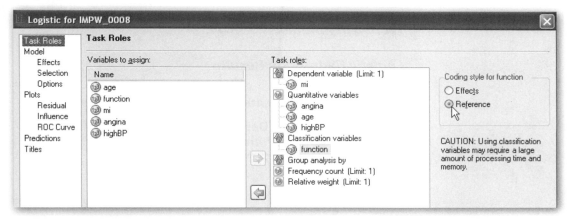

Table 7.3 Results from Logistic Regression of Coronary Data

Model Information		
Data Set	WORK.SORTTEMPTABLESORTED	
Response Variable	mi	mi
Number of Response Levels	2	
Model	binary logit	
Optimization Technique	Fisher's scoring	

Number of Observations Read	117
Number of Observations Used	117

Response Profile		
Ordered Value	mi	Total Frequency
1	1	74
2	0	43

Probability modeled is mi=1.

Logistic Regression Results

Class Level Information				
Class	Value	Design Variables		
function	0	1	0	0
	1	0	1	0
	2	0	0	1
	3	0	0	0

Model Convergence Status
Convergence criterion (GCONV=1E-8) satisfied.

Model Fit Statistics		
Criterion	Intercept Only	Intercept and Covariates
AIC	155.884	160.873
SC	158.646	180.208
-2 Log L	153.884	146.873

Testing Global Null Hypothesis: BETA=0			
Test	Chi-Square	DF	Pr > ChiSq
Likelihood Ratio	7.0110	6	0.3198
Score	6.5948	6	0.3599
Wald	6.1602	6	0.4055

Type 3 Analysis of Effects			
Effect	DF	Wald Chi-Square	Pr > ChiSq
age	1	0.5858	0.4441
angina	1	3.8335	0.0502
highBP	1	1.4023	0.2363
function	3	1.8464	0.6049

Analysis of Maximum Likelihood Estimates						
Parameter		DF	Estimate	Standard Error	Wald Chi-Square	Pr > ChiSq
Intercept		1	2.9311	1.4363	4.1644	0.0413
age		1	-0.0164	0.0214	0.5858	0.4441
angina		1	-1.2074	0.6167	3.8335	0.0502
highBP		1	-0.5090	0.4298	1.4023	0.2363
function	0	1	-0.7678	0.6377	1.4495	0.2286
function	1	1	-0.6114	0.6376	0.9193	0.3376
function	2	1	-0.2351	0.5310	0.1960	0.6580

Odds Ratio Estimates			
Effect	Point Estimate	95% Wald Confidence Limits	
age	0.984	0.943	1.026
angina	0.299	0.089	1.001
highBP	0.601	0.259	1.396
function 0 vs 3	0.464	0.133	1.619
function 1 vs 3	0.543	0.156	1.893
function 2 vs 3	0.791	0.279	2.238

Association of Predicted Probabilities and Observed Responses			
Percent Concordant	66.8	Somers' D	0.338
Percent Discordant	33.0	Gamma	0.339
Percent Tied	0.3	Tau-a	0.159
Pairs	3182	C	0.669

The first part of Table 7.3 to comment on is the **Class Level Information**, which shows how the four-category variable **function** has been coded in terms of three dummy variables (called *design variables* in Table 7.3). The first of the three dummy variables represents a comparison of **function**=0 and **function**=3; the second, a comparison of **function**=1 and **function**=3; and the third, a comparison of **function**=2 and **function**=3. The estimated regression coefficients for each dummy variable are interpreted in exactly the same way as explained previously for the first model fitted using only **angina**. So, for example, the exponentiated regression coefficient for the first dummy variable (0.464) gives the odds ratio for comparing categories 0 and 3 of the variable **function**. But if we look at the **Testing Global Null Hypothesis: BETA=0** section of Table 7.3, each of the three tests of the hypothesis that all the regression coefficients in the model are 0 suggest that the hypothesis should be accepted. So we have to conclude that none of the four explanatory variables have much effect on the occurrence of an infarct. (The model fitted considers only the linear effect of age. There may be a curvilinear effect; see Exercise 4.2 in Chapter 4.)

Although none of the regression coefficients can be claimed to be significant, we can use the coefficient for age to illustrate what was said earlier about interpreting the coefficient as a change in the log odds of the response variable when the associated explanatory variable changes by one unit. From Table 7.3, we see that an increase of one year in age decreases the log odds of an infarct by 0.0164, conditional on the other three explanatory variables remaining constant. (Forget for the moment that the regression coefficient for age is not significantly different from zero; it is not relevant here.) But suppose we were interested in the change associated with a 10-year increase in age? Such a change is simply 10 x (-0.0164)=-0.164. So a 10-year increase in age decreases the log odds of an infarct by 0.164. Using the estimated standard error of the regression coefficient for age from Table 7.3 (namely 0.0214), we can next calculate the 95% confidence interval for the change in log odds associated with a ten-year increase in age as [−0.164−1.96x10x0.0214,−0.164+1.96x10x0.0214], that is [−0.583,0.255]. We can exponentiate both the point estimate and the limits of the confidence interval to give the result for the odds ratio. Undertaking the calculation gives the estimated odds ratio as 0.849 and the 95% confidence interval as [0.558,1.290]. (As we would expect, the confidence interval contains the value 1 since we have already shown that there is no evidence of an age effect in determining past history of an infarct.)

To make the process described above simple to apply using SAS Enterprise Guide, the **Task Roles** pane has the option to specify the number of units for which the log odds and odds ratios are to be calculated. The specification can be either in the original units, such as years, or in terms of standard deviations. The standard deviation units can be useful for comparing the effects of continuous predictors that are measured on different scales. Display 7.3 shows how the results above could be produced.

Display 7.3 Task Roles Pane for Coronary Data Showing How to Calculate Results for 10 Years of Age

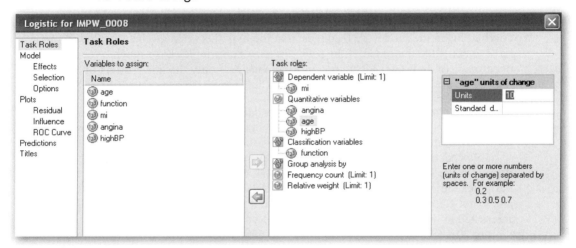

7.3 Exercises

Exercise 7.1 The data set **plasma** was collected to examine the extent to which erythrocyte sedimentation rate (ESR), i.e., the rate at which red blood cells (erythocytes) settle out of suspension in blood plasma, is related to two plasma proteins, fibrinogen and γ-globulin, both measured in gm/l. The ESR for a healthy individual should be less than 20mm/h. Since the absolute value of ESR is relatively unimportant, the response variable used here denotes whether or not this is the case. A response of 0 signifies a healthy individual (ESR<20) while a response of unity refers to an unhealthy individual (ESR≥20). The aim of the analysis for these data is to determine the strength of any relationship between the ESR level and the levels of the two plasmas. Investigate the relationship by fitting a logistic model for the probability of an unhealthy individual with fibrinogen and gamma as the two explanatory variables. What are your conclusions?

Fibrinogen	Gamma	ESR
2.52	38	0
2.56	31	0
2.19	33	0
2.18	31	0
3.41	37	0
2.46	36	0
3.22	38	0
2.21	37	0
3.15	39	0
2.60	41	0
2.29	36	0
2.35	29	0
5.06	37	1
3.34	32	1
2.38	37	1
3.15	36	0
3.53	46	1
2.68	34	0
2.60	38	0
2.23	37	0
2.88	30	0
2.65	46	0
2.09	44	1
2.28	36	0
2.67	39	0
2.29	31	0
2.15	31	0

(continued)

Fibrinogen	Gamma	ESR
2.54	28	0
3.93	32	1
3.34	30	0
2.99	36	0
3.32	35	0

Exercise 7.2 The data in **leukaemia2** show whether or not patients with leukemia lived for at least 24 weeks after diagnosis along with the values of two explanatory variables, the white blood count and the presence or absence of a morphological characteristic of the white blood cells (AG). The data are from Venables and Ripley (1994). Fit a logistic regression model to the data to determine whether the explanatory variables are predictive of survival longer than 24 weeks. (You may need to consider an interaction term.)

White blood count	AG	Survival longer than 24 weeks from diagnosis (1=yes,0=no)
2300	present	1
750	present	1
4300	present	1
2600	present	1
6000	present	0
10500	present	1
10000	present	1
17000	present	0
5400	present	1
7000	present	1
9400	present	1
32000	present	1
35000	present	0
100000	present	0
100000	present	0

(*continued*)

White blood count	AG	Survival longer than 24 weeks from diagnosis (1=yes,0=no)
52000	present	0
100000	present	1
4400	absent	1
3000	absent	1
4000	absent	0
1500	absent	0
9000	absent	0
5300	absent	0
10000	absent	0
19000	absent	0
27000	absent	0
28000	absent	0
31000	absent	0
26000	absent	0
21000	absent	0
79000	absent	1
100000	absent	0
100000	absent	1

Exercise 7.3 The data in **lowbwgt** comprise part of the data set given in Hosmer and Lemeshow (1989), which was collected during a study to identify risk factors associated with giving birth to a low birthweight baby, defined as weighing less than 2500 gms. The risk factors considered were age of the mother, weight of the mother at her last menstrual period, race of mother, and number of physician visits during the first trimester of the pregnancy. Fit a logistic regression model for the probability of a low birthweight infant using age, lwt, race (coded in terms of two dummy variables) and ftv as explanatory variables. What conclusions do you draw from the fitted model?

Low Infant Birthweight Data

ID	LOW	AGE	LWT	RACE	FTV
85	0	19	182	2	0
86	0	33	155	3	3
87	0	20	105	1	1
88	0	21	108	1	2
. . . .					
79	1	28	95	1	2
81	1	14	100	3	2
82	1	23	94	3	0
83	1	17	142	2	0
84	1	21	130	1	3

LOW	:	0 = weight of baby >2500g
		1 = weight of baby <= 2500g
AGE	:	Age of mother in years
LWT	:	weight of mother at last menstrual period
RACE	:	1 = white, 2 = black, 3 = other
FTV	:	Number of physician visits in the first trimester

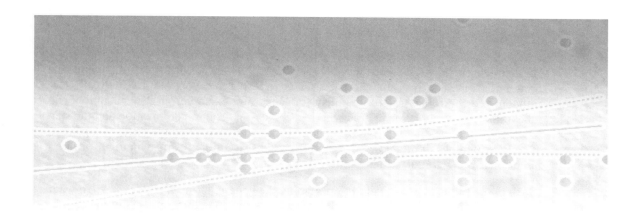

Chapter 8

Survival Analysis

8.1 Introduction

In many research studies, particularly in medicine but also in other areas, the main outcome variable is the *time* to the occurrence of a specific event of interest. In a randomized controlled trial of treatment for cancer, for example, surgery, radiation, and chemotherapy might be compared with respect to time from randomization and the start of therapy until death. In such a trial, the event of interest is often the death of a patient. In other situations, it might be remission from a disease, relief from symptoms, or the recurrence of a particular condition.

Such observations are generally referred to by the generic term *survival data* even when the endpoint or event being considered is not death but something else. Survival data often require special approaches to their analyses for two main reasons:

- Survival data are generally not symmetrically distributed. They will often be positively skewed (frequently to a considerable degree), with a few people surviving a long time compared with the majority. Consequently, an assumption or expectation of normality is, *a prior*, very likely to be quite unrealistic.

- At the completion of the study, some patients may not have reached the endpoint of interest (death, relapse, etc.); in such cases, the exact survival times are not known, although we do know that the survival times are greater than the length of time the individual has been in the study. The survival times of such individuals are said to be *censored* (more precisely, they are *right-censored*).

In this chapter, we will be concerned with how to deal with survival data. The statistical topics we will cover are:

- Observations for which the event of interest has not occurred when the study ends: Censored observations

- Methods for describing the survival experience of the individuals in the sample: Survival and hazard functions

- Estimating survival functions using the Kaplan-Meier approach

- Assessing the effects of explanatory variables on survival: Cox regression

8.2 Example: Gastric Cancer

The data in Table 8.1 show the number of days from diagnosis to either death or the end of the study of two groups of 45 patients suffering from gastric cancer. Group 1 received chemotherapy and radiation; Group 2 received only chemotherapy. For the patients who died, we have their *survival times* but at the end of the study some patients remained alive so their survival times are unknown. But we do know that the survival times for such patients are *longer* than the number of days given in Table 8.1. Such observations are denoted as having been *censored* and are indicated by asterisks in Table 8.1. The question of interest is whether the data provide any evidence that patients survive longer under one treatment regimen than under the other?

Table 8.1 Survival Times for Patients with Gastric Cancer, Under Two
Different Treatment Regimes

```
Group 1
17,42,44,48,60,72,74,95,103,108,122,144,167,170,183,185,193,195,19
7,208,234,235,254,307,315,401,445,464,484,528,542,567,577,580,795,
855,1174*,1214,1232*,1366,1455*,1585*,1622*,1626*,1936*
```

```
Group 2
1,63,105,125,182,216,250,262,301,301,342,354,356,358,380,383,383,
388, 394, 408, 460, 489, 523, 524, 535, 562, 569, 675, 676, 748,
778, 786,797, 955, 968, 977, 1245, 1271, 1420, 1460*, 1516*,
1551,1690*, 1694
```

8.2.1 Gastric Cancer Patients: Summarizing and Displaying Their Survival Experience Using the Survival Function

Of central importance in the analysis of survival data are two functions used to describe the distribution of the data, the *survival function*, and the *hazard function*. For the moment, we shall concentrate on the survival function and return to the hazard function in Section 8.3.

In essence, the survival function is a very simple plot of the proportion of subjects surviving for at least time t plotted against t. When there are no censored observations in the data, the survival function can be found in a straightforward way since the proportions required to construct the plot are calculated simply as number of individuals with survival times greater than equal to t divided by the number of individuals in the data set. Since every patient is alive at the beginning of the study and no one is observed to survive longer than the largest survival time, the survival function for $t=0$ is one and for $t=$maximum observed value the survival function is zero.

But a quintessential part of the vast majority of survival time data sets, like those in Section 8.2, is the presence of censored observations, in which case the simple approach to finding the survival function described above is not appropriate. To estimate the survival function for data containing censored observations, the most usual method is that described in Kaplan and Meier (1958) and known generally as the *product-limit estimator* (or *Kaplan-Meier estimator*). The essence of the procedure is the use of the continued product of a series of conditional probabilities, but we shall not give details here but instead refer you to Everitt (1994) and Der and Everitt (2005).

8.2.2 Plotting Survival Functions Using SAS Enterprise Guide

Plotting the estimated survival functions of two groups of survival times on the same diagram allows us an informal comparison of the survival experience of the two (or, in some examples, more than two) groups. We can illustrate the construction of survival functions using SAS Enterprise Guide on the survival times for gastric cancer in Table 8.1. The data are in a SAS data set, **stomachca**. Add them to the project:

1. Select **File≻Open≻Data≻Local Computer**, browse to the location **c:\saseg\sasdata**, and click **Open**.

Now, use the **Life Tables** task to construct the plot:

1. Select **Analyze≻Survival Analysis≻Life Tables**.

2. Under **Task Roles**, assign **days** the role of **Survival time**, **censor** as the **Censoring variable**, and select **1** as the **Right censoring value** (Display 8.1). Treat **group** as a **Strata variable**.

3. Under **Method**, the default estimation method is **Product-limit** (**Kaplan-Meier**), so that can be left as it is.

4. Under **Plots**, select **Show survival function plot**, **Show censored values**, and **Overlay strata on a single plot**.

5. Click **Run**.

Running the task gives Figure 8.1 along with the output shown in Table 8.2.

Display 8.1 Task Roles Pane for Life Tables Analysis of Stomach Cancer
Data

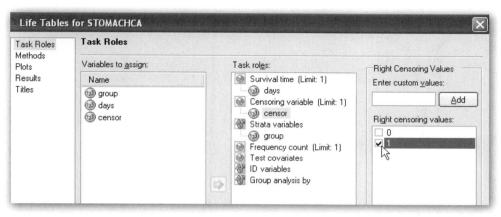

The estimated survival functions in Figure 8.1 suggest that survival is longer under the
second treatment regimen, chemotherapy only. The details of the product-limit estimation
of each survival function are given in Table 8.2, but of more interest are the quartile
estimates of survival given for each treatment group. In particular, the estimated median
survival times of 254 days (95% confidence interval, [193,484]) for Group 1 and 506
days (95% confidence interval [383,676]) for Group 2. Median survival time is estimated
to be approximately twice as long for chemotherapy alone as for radiation plus
chemotherapy, although the confidence intervals are wide.

Figure 8.1 Estimated Survival Functions for Two Treatment Regimes for Gastric Cancer

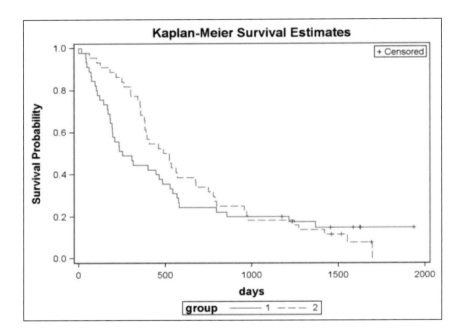

Table 8.2 Details of Estimation of Survival Functions for Gastric Cancer Data and Test of the Equality of the Two Survival Functions

	Stratum 1: group = 1				
	Product-Limit Survival Estimates				
days	Survival	Failure	Survival Standard Error	Number Failed	Number Left
0.00	1.0000	0	0	0	45
17.00	0.9778	0.0222	0.0220	1	44
42.00	0.9556	0.0444	0.0307	2	43
44.00	0.9333	0.0667	0.0372	3	42
48.00	0.9111	0.0889	0.0424	4	41
60.00	0.8889	0.1111	0.0468	5	40
72.00	0.8667	0.1333	0.0507	6	39
74.00	0.8444	0.1556	0.0540	7	38
95.00	0.8222	0.1778	0.0570	8	37
103.00	0.8000	0.2000	0.0596	9	36
108.00	0.7778	0.2222	0.0620	10	35
122.00	0.7556	0.2444	0.0641	11	34
144.00	0.7333	0.2667	0.0659	12	33
167.00	0.7111	0.2889	0.0676	13	32
170.00	0.6889	0.3111	0.0690	14	31
183.00	0.6667	0.3333	0.0703	15	30
185.00	0.6444	0.3556	0.0714	16	29
193.00	0.6222	0.3778	0.0723	17	28
195.00	0.6000	0.4000	0.0730	18	27
197.00	0.5778	0.4222	0.0736	19	26

		Stratum 1: group = 1			
		Product-Limit Survival Estimates			
days	Survival	Failure	Survival Standard Error	Number Failed	Number Left
208.00	0.5556	0.4444	0.0741	20	25
234.00	0.5333	0.4667	0.0744	21	24
235.00	0.5111	0.4889	0.0745	22	23
254.00	0.4889	0.5111	0.0745	23	22
307.00	0.4667	0.5333	0.0744	24	21
315.00	0.4444	0.5556	0.0741	25	20
401.00	0.4222	0.5778	0.0736	26	19
445.00	0.4000	0.6000	0.0730	27	18
464.00	0.3778	0.6222	0.0723	28	17
484.00	0.3556	0.6444	0.0714	29	16
528.00	0.3333	0.6667	0.0703	30	15
542.00	0.3111	0.6889	0.0690	31	14
567.00	0.2889	0.7111	0.0676	32	13
577.00	0.2667	0.7333	0.0659	33	12
580.00	0.2444	0.7556	0.0641	34	11
795.00	0.2222	0.7778	0.0620	35	10
855.00	0.2000	0.8000	0.0596	36	9
1174.00 *	.	.	.	36	8
1214.00	0.1750	0.8250	0.0572	37	7
1232.00 *	.	.	.	37	6
1366.00	0.1458	0.8542	0.0546	38	5
1455.00 *	.	.	.	38	4
1585.00 *	.	.	.	38	3

		Stratum 1: group = 1			
		Product-Limit Survival Estimates			
days	Survival	Failure	Survival Standard Error	Number Failed	Number Left
1622.00 *	.	.	.	38	2
1626.00 *	.	.	.	38	1
1936.00 *	.	.	.	38	0

The marked survival times are censored observations.

Summary Statistics for Time Variable Days

	Quartile Estimates		
		95% Confidence Interval	
Percent	Point Estimate	[Lower	Upper)
75	580.00	464.00	.
50	254.00	193.00	484.00
25	144.00	74.00	195.00

Mean	Standard Error
491.84	71.01

The mean survival time and its standard error were underestimated because the largest observation was censored and the estimation was restricted to the largest event time.

			Stratum 2: group = 2		
			Product-Limit Survival Estimates		
days	**Survival**	**Failure**	**Survival Standard Error**	**Number Failed**	**Number Left**
0.00	1.0000	0	0	0	44
1.00	0.9773	0.0227	0.0225	1	43
63.00	0.9545	0.0455	0.0314	2	42
105.00	0.9318	0.0682	0.0380	3	41
125.00	0.9091	0.0909	0.0433	4	40
182.00	0.8864	0.1136	0.0478	5	39
216.00	0.8636	0.1364	0.0517	6	38
250.00	0.8409	0.1591	0.0551	7	37
262.00	0.8182	0.1818	0.0581	8	36
301.00	.	.	.	9	35
301.00	0.7727	0.2273	0.0632	10	34
342.00	0.7500	0.2500	0.0653	11	33
354.00	0.7273	0.2727	0.0671	12	32
356.00	0.7045	0.2955	0.0688	13	31
358.00	0.6818	0.3182	0.0702	14	30
380.00	0.6591	0.3409	0.0715	15	29
383.00	.	.	.	16	28
383.00	0.6136	0.3864	0.0734	17	27
388.00	0.5909	0.4091	0.0741	18	26
394.00	0.5682	0.4318	0.0747	19	25
408.00	0.5455	0.4545	0.0751	20	24
460.00	0.5227	0.4773	0.0753	21	23
489.00	0.5000	0.5000	0.0754	22	22

days	Survival	Failure	Survival Standard Error	Number Failed	Number Left
	Stratum 2: group = 2				
	Product-Limit Survival Estimates				
523.00	0.4773	0.5227	0.0753	23	21
524.00	0.4545	0.5455	0.0751	24	20
535.00	0.4318	0.5682	0.0747	25	19
562.00	0.4091	0.5909	0.0741	26	18
569.00	0.3864	0.6136	0.0734	27	17
675.00	0.3636	0.6364	0.0725	28	16
676.00	0.3409	0.6591	0.0715	29	15
748.00	0.3182	0.6818	0.0702	30	14
778.00	0.2955	0.7045	0.0688	31	13
786.00	0.2727	0.7273	0.0671	32	12
797.00	0.2500	0.7500	0.0653	33	11
955.00	0.2273	0.7727	0.0632	34	10
968.00	0.2045	0.7955	0.0608	35	9
977.00	0.1818	0.8182	0.0581	36	8
1245.00	0.1591	0.8409	0.0551	37	7
1271.00	0.1364	0.8636	0.0517	38	6
1420.00	0.1136	0.8864	0.0478	39	5
1460.00 *	.	.	.	39	4
1516.00 *	.	.	.	39	3
1551.00	0.0758	0.9242	0.0444	40	2
1690.00 *	.	.	.	40	1
1694.00	0	1.0000	0	41	0

The marked survival times are censored observations.

Summary Statistics for Time Variable Days

	Quartile Estimates			
			95% Confidence Interval	
Percent		Point Estimate	[Lower	Upper)
75		876.00	569.00	1271.00
50		506.00	383.00	676.00
25		348.00	250.00	388.00

Mean	Standard Error
653.22	72.35

Summary of the Number of Censored and Uncensored Values

Stratum	group	Total	Failed	Censored	Percent Censored
1	1	45	38	7	15.56
2	2	44	41	3	6.82
Total		89	79	10	11.24

8.2.3 Testing the Equality of Two Survival Functions: The Log-Rank Test

Graphical examination of the two survival functions for the gastric cancer patients is a helpful and essential initial step in the analysis of the data. But is there a way to make a more formal comparison of the two survival functions? In other words, how can we test the following null hypothesis that the two survival functions are the same?

$$H_0 : S_1 = S_2$$

where S_1 and S_2 are, respectively, the population survival functions of groups 1 and 2. In the absence of censored observations, standard nonparametric tests such as the Wilcoxon-Mann-Whitney test might be used to compare the survival times of each group (see Chapter 2). When there are censored observations, there are a number of tests, both parametric and nonparametric, that might be used to compare groups of survival times. Most common is the *log-rank test* which compares the observed number of "deaths" occurring at each particular time point with the number to be expected if the survival experience of the two groups is the same. Full details of the log-rank test are given in Everitt (1994).

To assess the survival experience of the two treatment regimes more formally, we have to look at the **Test of Equality over Strata** section of the output from the previous SAS Enterprise Guide instructions for calculating the survival functions. The output is given here in Table 8.3. Three tests of the null hypothesis that there is no difference in the survivor functions for the two treatments groups are given, one being the log-rank test mentioned previously, and the other two being the Wilcoxon test and a likelihood ratio test. Details of all three tests can be found in Collett (2002). Here, two of the tests suggest that there is no evidence against the null hypothesis and that the two treatments do not lead to different survival experiences for patients.

But the result from the Wilcoxon test is quite different with an associated *p*-value of 0.0378 indicating some evidence against the null hypothesis. The reason for the difference is that the log-rank test (and the likelihood ratio test which assumes that the distribution of survival times is exponential) are most useful when the population survival functions of the two groups do not cross, indicating that the hazard functions of the two groups are proportional (see Section 8.3.1). Here, the survival functions estimated from the sample observations do cross, suggesting perhaps that the population functions might also cross. When there is a crossing of survival functions, the Wilcoxon test is more sensitive to differences between groups in the early time points, and the log-rank test is more sensitive to later differences. Consequently, it might be legitimate here to claim that there is evidence that survival using chemotherapy only is longer than with radiation plus chemotherapy particularly early in the course of treatment.

Table 8.3 Tests for Equality of Survivor Functions for Gastric Cancer Patients Given Two Different Treatments

Testing Homogeneity of Survival Curves for Days over Strata

Test	Chi-Square	DF	Pr > Chi-Square
Test of Equality over Strata			
Log-Rank	0.5654	1	0.4521
Wilcoxon	4.3162	1	0.0378
-2Log(LR)	0.3130	1	0.5758

8.3 Example: Myeloblastic Leukemia

The data in Table 8.4 give the survival times in months of 51 adult patients with acute myeloblastic leukemia along with the values of five other variables that may or may not affect survival time. Here, the aim will be to construct a suitable statistical model that will allow us to say which of the five explanatory variables are of greatest importance in determining survival time in patients suffering from leukemia.

Table 8.4 Data for 51 Leukemia Patients

Variable						
1	2	3	4	5	6	7
20	78	39	7	990	18	0
25	64	61	16	1030	31	1
26	61	55	12	982	31	0
26	64	64	16	100	31	0
27	95	95	6	980	36	0
27	80	64	8	1010	1	0
28	88	88	20	986	9	0
28	70	70	14	1010	39	1
31	72	72	5	988	20	1
33	58	58	7	986	4	0
33	92	92	5	980	45	1
33	42	38	12	984	36	0
34	26	26	7	982	12	0
36	55	55	14	986	8	0
37	71	71	15	1020	1	0
40	91	91	9	986	15	0
40	52	49	12	988	24	0
43	74	63	4	986	2	0
45	78	47	14	980	33	0
45	60	36	10	992	29	1
45	82	32	10	1016	7	0
45	79	79	4	1030	0	0
47	56	28	2	990	1	0
48	60	54	10	1002	2	0
50	83	66	19	996	12	0
50	36	32	14	992	9	0
51	88	70	8	982	1	0
52	87	87	7	986	1	0

(continued)

Table 8.4 (*continued*)

Variable						
1	**2**	**3**	**4**	**5**	**6**	**7**
53	75	68	13	980	9	0
53	65	65	6	982	5	0
56	97	92	10	992	27	1
57	87	83	19	1020	1	0
59	45	45	8	999	13	0
59	36	34	5	1038	1	0
60	39	33	7	988	5	0
60	76	53	12	982	1	0
61	46	37	4	1006	3	0
61	39	8	8	990	4	0
61	90	90	11	990	1	0
62	84	84	19	1020	18	0
63	42	27	5	1014	1	0
65	75	75	10	1004	2	0
71	44	22	6	990	1	0
71	63	63	11	986	8	0
73	33	33	4	1010	3	0
73	93	84	6	1020	4	0
74	58	58	10	1002	14	0
74	32	30	16	988	3	0
75	60	60	17	990	13	0
77	69	69	9	986	13	0
80	73	73	7	986	1	0

Variables

1	Age at diagnosis
2	Smear differential percentage of blasts
3	Percentage of absolute marrow leukemia infiltrate
4	Percentage labeling index of the bone marrow leukemia cells
5	Highest temperature prior to treatment (degrees F. decimal points omitted)
6	Survival time from diagnosis (months)
7	Status at end of study(0=dead, 1=alive)

8.3.1 What Affects Survival in Patients with Leukemia? The Hazard Function and Cox Regression

For the leukemia data in Table 8.4, the main question of interest is which of the five explanatory or *prognostic* variables are of most importance in predicting a patient's survival time? The same question is posed in Chapter 6 for continuous response variables leading to multiple linear regression, and in Chapter 7 for binary response variables leading to logistic regression.

But neither multiple regression nor logistic regression is suitable for modeling survival time data because of the special features of such data, in particular the censoring that almost always occurs. A number of more suitable models have therefore been developed, of which the most successful (certainly the most widely used) is that due to Cox (1972). But before describing the method, we now need to say a little more about the hazard function mentioned in passing in Section 8.2.

The Hazard Function

In dealing with survival time data, it is often of great interest to assess which periods have the highest and which the lowest chances of death (or whatever the event of interest may be) amongst those still alive (and therefore at risk) at the time. The appropriate approach to assessing such risks is the *hazard function* which is defined as the instantaneous risk that an individual dies (or experiences the event of interest) in a small time interval, given that the individual has survived up to the beginning of the interval. The hazard function is also known as the *instantaneous failure rate*, the *instantaneous death rate* and the *age-specific failure rate*. It is a measure of how likely an individual is to die as a function of the age of the individual.

The conditioning feature of the definition of the hazard function is of central importance. For example, the probability of dying at age 100 is very small since most people die before that age. In contrast, the probability of a person dying at age 100, having reached that age, is much greater. The hazard function may remain constant, increase, decrease, or take on some more complex shape. The hazard function for death in human beings, for example, has approximately the shape shown in Figure 8.2. The hazard function is relatively high immediately after birth, declines rapidly in the early years of life, remains almost constant during middle age, and then begins to rise again in old age.

Figure 8.2 Hazard Function for Death in Humans

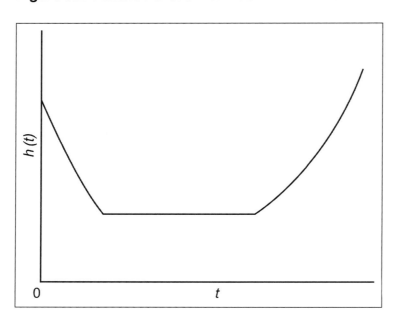

Cox Regression

Having described the hazard function, we can now move on to consider *Cox regression* for assessing how some prognostic variables of interest affect survival times. The essential feature of Cox regression is the modeling of the hazard function which provides a simpler vehicle for assessing the joint effects of prognostic variables than the survival function since it does not involve the cumulative history of events. Since the hazard function is restricted to being positive, a possible model is:

$$\log[h(t)] = \beta_0 + \beta_1 x_1 + \beta_2 x_2 + ... + \beta_p x_p$$

Where $x_1, x_2, ..., x_p$ are the explanatory variables and $h(t)$ is the hazard function. This would be a suitable model only for the hazard function that is constant over time. Such a model is very restrictive since hazards that increase or decrease with time, or have some more complex form, are far more likely to occur in practice. But it may be difficult to find the appropriate explicit function of time to include in the model above and, rather than trying, Cox regression finesses the problem by introducing an *arbitrary baseline* hazard function into the model to give

$$\log[h(t)] = \log[h_0(t)] + \beta_1 x_1 + \beta_2 x_2 + \dots + \beta_p x_p$$

The baseline hazard function, $h_0(t)$, is left unspecified, and the model forces the hazard ratio of two individuals to be constant over time; so if an individual has a risk of death at some initial time point that is twice as high as that of some other individual, then the risk of death remains twice as high at all later time points. Hence, the term *proportional hazards model* is an alternative name for Cox regression. The proportional hazard aspect of the model and how the parameters in the model are estimated are detailed in Der and Everitt (2005) and Everitt and Rabe-Hesketh (2001). Here, we will concentrate on how we can fit the model to data using SAS Enterprise Guide and how to interpret the parameter estimates which result from the fitting process.

8.3.2 Applying Cox Regression Using SAS Enterprise Guide

To apply Cox regression to the leukemia data in Table 8.4, we begin by opening a new process flow window for the analysis.

1. Select **File≻New≻Process Flow**.

2. Rename the process flow **Leukemia** (right-click on the **Process Flow** tab and select **Rename**.

The data are already available in a SAS data set and so can be simply added to the project.

1. Select **File≻Open≻Data≻Local Computer**, browse to the location of the data set **c:\saseg\sasdata**, select **leukaemia.sas7bdat**, and **Open**.

For the analysis:

1. Select **Analyze≻Survival Analysis≻Proportional Hazards**.

2. Under **Task Roles**, assign **months** as the **Survival time** and **status** as the **Censoring variable**, select **1** as the **Right censoring value** (Display 8.2) and assign all the remaining variables as **Explanatory variables**. It is worth noting that the **Proportional Hazards** task does not have the option of including classification variables, so these would have to be recoded as a series of dummy variables in order to be included.

3. Under **Model**, the default model includes all explanatory variables.

4. Under **Methods**, select **Compute confidence limits for hazard ratios.**

5. Click **Run**.

The results are given in Table 8.5.

Display 8.2 Task Roles Pane for Cox Regression of the Leukemia Data

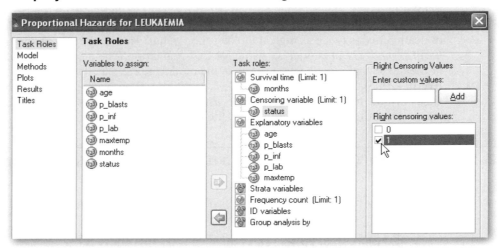

First, the tests that all the regression coefficients in the Cox regression model for the leukemia data are zero given in the **Testing Global Null Hypothesis** section of Table 8.5 all have associated *p*-values less than 0.05 so there is evidence that at least some of the regression coefficients differ from zero. Moving on to the **Analysis of Maximum Likelihood Estimates** section of Table 8.5, comparing each estimated coefficient with its estimated standard error suggests that age at diagnosis alone is an important prognostic variable for survival time. The estimated regression coefficient for age is 0.03359 with a standard error of 0.01036. The associated *p*-value is 0.0012. The estimated regression coefficient for age at diagnosis is interpreted in much the same way as are the regression coefficients in multiple regression and logistic regression. For the leukemia data, an increase in one year for age at diagnosis increases the logarithm of the hazard function by 0.0336.

A more appealing interpretation results if the regression coefficient is exponentiated to give the value 1.034 as also shown in the **Analysis of Maximum Likelihood Estimates** section of Table 8.5. The value 1.034 implies that the hazard function of an individual aged *x*+1 at diagnosis is 1.034 times the hazard function of an individual whose age at diagnosis is *x*. The corresponding 95% confidence interval also given in Table 8.5 is [1.013,1.055].

An additional aid to interpretation is found by first calculating 100(exp(coefficient)-1) which gives the percentage change in the hazard function with each unit change in the explanatory variable. Applying the calculation to the estimated regression coefficient for age at diagnosis, we conclude that a yearly increase in age at diagnosis leads to an estimated 3.4% increase in the hazard function, with 95% confidence limits [1.3%,5.5%].

Table 8.5 Results of Applying Cox Regression to the Leukemia Data in Table 8.2

Model Information	
Data Set	WORK.TMP0TEMPTABLEINPUT
Dependent Variable	Months
Censoring Variable	Status
Censoring Value(s)	1
Ties Handling	BRESLOW

Number of Observations Read	51
Number of Observations Used	51

Summary of the Number of Event and Censored Values			
Total	Event	Censored	Percent Censored
51	45	6	11.76

Convergence Status
Convergence criterion (GCONV=1E-8) satisfied.

Model Fit Statistics		
Criterion	Without Covariates	With Covariates
-2 LOG L	291.106	276.086
AIC	291.106	286.086
SBC	291.106	295.120

Testing Global Null Hypothesis: BETA=0			
Test	Chi-Square	DF	Pr > ChiSq
Likelihood Ratio	15.0194	5	0.0103
Score	14.8274	5	0.0111
Wald	14.0130	5	0.0155

Analysis of Maximum Likelihood Estimates								
Variable	DF	Parameter Estimate	Standard Error	Chi-Square	Pr > ChiSq	Hazard Ratio	95% Hazard Ratio Confidence Limits	
age	1	0.03359	0.01036	10.5140	0.0012	1.034	1.013	1.055
p_blasts	1	0.00928	0.01473	0.3968	0.5287	1.009	0.981	1.039
p_inf	1	-0.01613	0.01267	1.6195	0.2032	0.984	0.960	1.009
p_lab	1	-0.05386	0.03899	1.9086	0.1671	0.948	0.878	1.023
maxtemp	1	-0.0003663	0.00128	0.0820	0.7746	1.000	0.997	1.002

8.4 Exercises

Exercise 8.1 The **breast** data set gives the survival times after mastectomy of women with breast cancer. Based on a histochemical marker, the cancers were classified as having metastasized or not. Censoring is indicated by an asterisk. Plot the product-limit estimates of the two survival functions on the same diagram, find the median survival times, and test for any difference in the survival experience of the two groups of women.

```
Not metastasized
23 47 69 70* 71* 100* 101* 148 181 198* 208* 212* 224*

Metastasized
5 8 10 13 18 24 26 26 31 35 40 41 48 50 59 61 68 71 76* 105* 107*
109* 113 116* 118 143 154* 162* 188* 212* 217* 225*
```

Exercise 8.2 The data in **prostate** arise from a randomized controlled trial to compare two treatments for prostrate cancer. Patients were randomized to receive either 1mg of diethylstilbestrol (DES) or 1mg of placebo daily by mouth, and their survival was recorded in months. The variables in the table below are as follows:

Treatment	0 = placebo, 1 = 1mg of diethylstillbeterol daily
Status	1 = dead, 0 = censored
Time	Survival time in months
Age	Age at trial entry in years
Haem	Serum haemoglobin level in gm/100ml
Size	Size of primary tumor in centimeters squared
Gleason	the value of a combined index of tumor stage and grade (the larger the index, the more advanced the tumor)

Fit a Cox regression to the data and identify the most important prognostic variables for survival.

Prostate Cancer Trial Data

Treatment	Time	Status	Age	Haem	Size	Gleason
0	65	0	67	13.4	34	8
1	61	0	60	14.4	4	10
1	60	0	77	15.6	3	8
0	58	0	64	16.21	6	9
1	51	0	65	14.1	21	9
0	14	1	73	12.4	18	11
0	43	0	60	13.6	7	9
1	16	0	73	13.8	8	9
0	52	0	73	11.7	5	9
. . . .						
1	67	0	73	13.8	7	8
0	23	0	68	12.5	2	8
0	62	0	63	13.2	3	8

Exercise 8.3 The data set **Heroin** gives the times that heroin addicts remained in a clinic for methadone treatment. If they were still in treatment at the end of the study, the status variable has a value 0. Potential explanatory variables for time to complete treatment are maximum methadone dose, clinic where treatment took place, and whether or not the addict had a criminal record; yes is coded one and no coded zero.

ID	Clinic	Status	Time	Prison	Dose	ID	Clinic	Status	Time	Prison	Dose
1	1	1	428	0	50	132	2	0	633	0	70
2	1	1	275	1	55	133	2	1	661	0	40
3	1	1	262	0	55	134	2	1	232	1	70
4	1	1	183	0	30	135	2	1	13	1	60
5	1	1	259	1	65	137	2	0	563	0	70
6	1	1	714	0	55	138	2	0	969	0	80
. . . .											
127	2	1	26	0	40	262	2	1	540	0	80
128	2	0	72	1	40	263	2	0	551	0	65
129	2	0	641	0	70	264	1	1	90	0	40
131	2	0	367	0	70	266	1	1	47	0	45

References

Agresti, A. 1996. *Introduction to Categorical Data Analysis*. New York: Wiley.

Aitkin, M. 1978. "The analysis of unbalanced cross classifications." Journal of the Royal Statistical Society Series, Series A, Vol. 141, No. 2: 195–223.

Altman, D. G. 1991. *Practical Statistics for Medical Research*. 2d ed. London: CRC/Chapman and Hall.

Cleveland, W. S. 1994. *The Elements of Graphing Data*. Murray Hill, NJ: Hobart Press.

Collett, D. 2002. *Modelling Binary Data*. 2d ed. London: Chapman and Hall/CRC.

Collett, D. 2003. *Modelling Survival Data in Medical Research*. 2d ed. London: Chapman and Hall/CRC.

Cook, R. D. and Weisberg, S. 1982. *Residuals and Influence in Regression*. London: CRC/Chapman and Hall.

Cox, D. R. 1972. "Regression models and life tables." Journal of the Royal Statistical Society, Series B, Vol. 34, No. 2: 187–200.

Der, G. and Everitt, B. S. 2005. *Statistical Analysis of Medical Data Using SAS*. London: Chapman and Hall/CRC.

Everitt, B. S. 1992. *The Analysis of Contingency Tables*. 2d ed. London: CRC/Chapman and Hall.

Everitt, B. S. 1994. *Statistical Methods for Medical Investigations*. 2d ed. London: Arnold.

Everitt, B. S. 1996. *Making Sense of Statistics in Psychology: A Second-Level Course*. Oxford: Oxford University Press.

Everitt, B. S. and Palmer, C. R. 2006. *Encyclopaedic Companion to Medical Statistics*. Arnold, London.

Everitt, B. S. and Rabe-Hesketh, S. 2001. *Analyzing Medical Data Using S-PLUS*. New York: Springer-Verlag.

Fisher, R. 1925. *Statistical Methods for Research Workers*. Edinburgh: Oliver and Boyd.

Hand, D. J., Daly, F., Lunn, D., McConway, K., and Ostrowski, E. 1993. *A Handbook of Small Datasets*. London: Chapman and Hall/CRC.

Howell, D. C. (1992). *Statistical Methods for Psychologists*. 3d ed. Belmont, CA: Duxbury Press.

Kaplan, E. L. and Meier, P. 1958. "Nonparametric estimation from incomplete observations." Journal of the American Statistical Association, 53, No. 282: 457–481.

Kapor, M. 1981. "Efficiency on ergocycle in relation to knee-joint angle and drag." Unpublished master's dissertation, University of Delhi.

Kasser, I. and Bruce, R. A. 1969. "Comparative Effects of Aging and Coronary Heart Disease on Submaximal and Maximal Exercise." Circulation, 39, 759–774.

Mallows, C. L. 1973. "Some comments on Cp." Technometrics 15, No. 4: 661–675.

Mann, L. 1981. "The baiting crowds in episodes of threatened suicide." Journal of Personality and Social Psychology, 41, 703–709.

Maxwell, S. E. and Delaney, H. D. 1990. *Designing Experiments and Analyzing Data.* Belmont, CA: Wadsworth.

Mehta, C. R. and Patel, N. R. 1986. "A hybrid algorithm for Fisher's exact test on unordered r × c contingency tables." Communications in Statistics 15(2): 387–403.

Nelder, J. A. 1977. "A reformulation of linear models." Journal of the Royal Statistical Society Series A, Vol. 140, No. 1: 48–77.

Rawlings, J. O., Sastry G. P. and Dickey, D. A. 2001. *Applied Regression Analysis: A Research Tool.* New York: Springer-Verlag.

Rickman, R., Mitchell, N., Dingman, J., and Dalen, J. E. 1974. "Changes in serum cholesterol during the Stillman diet." Journal of the American Medical Association, 228, Issue 1: 54–58.

Scheffe, H. 1953. "A method for judging all contrasts in the analysis of variance." Biometrika 40 (1-2): 87–110.

Tufte, E. R. 1983. *The Visual Display of Quantitative Information.* Cheshire, CT: Graphics Press.

Venables, W. N. and Ripley, B. D. 1994. *Modern Applied Statistics with S-Plus.* New York: Springer-Verlag.

Wetherill, G. B. 1982. *Elementary Statistical Methods.* 3d ed. London: Chapman and Hall/CRC.

Woodley, W. L., Simpson, J., Biondini, R., and Berkeley, J. 1977. "Rainfall results 1970–1975: Florida area cumulus experiment." Science, 195, No. 4280: 735–742.

Index

Books Available from SAS Press

Advanced Log-Linear Models Using SAS®
by **Daniel Zelterman**

Analysis of Clinical Trials Using SAS®: A Practical Guide
by **Alex Dmitrienko, Geert Molenberghs, Walter Offen, and Christy Chuang-Stein**

Analyzing Receiver Operating Characteristic Curves with SAS®
by **Mithat Gönen**

Annotate: Simply the Basics
by **Art Carpenter**

Applied Multivariate Statistics with SAS® Software, Second Edition
by **Ravindra Khattree**
and **Dayanand N. Naik**

Applied Statistics and the SAS® Programming Language, Fifth Edition
by **Ronald P. Cody**
and **Jeffrey K. Smith**

An Array of Challenges — Test Your SAS® Skills
by **Robert Virgile**

Basic Statistics Using SAS® Enterprise Guide®: A Primer
by **Geoff Der**
and **Brian S. Everitt**

Building Web Applications with SAS/IntrNet®: A Guide to the Application Dispatcher
by **Don Henderson**

Carpenter's Complete Guide to the SAS® Macro Language, Second Edition
by **Art Carpenter**

Carpenter's Complete Guide to the SAS® REPORT Procedure
by **Art Carpenter**

The Cartoon Guide to Statistics
by **Larry Gonick**
and **Woollcott Smith**

Categorical Data Analysis Using the SAS® System, Second Edition
by **Maura E. Stokes, Charles S. Davis, and Gary G. Koch**

Cody's Data Cleaning Techniques Using SAS® Software
by **Ron Cody**

Common Statistical Methods for Clinical Research with SAS® Examples, Second Edition
by **Glenn A. Walker**

The Complete Guide to SAS® Indexes
by **Michael A. Raithel**

CRM Segmemtation and Clustering Using SAS® Enterprise Miner™
by **Randall S. Collica**

Data Management and Reporting Made Easy with SAS® Learning Edition 2.0
by **Sunil K. Gupta**

Data Preparation for Analytics Using SAS®
by **Gerhard Svolba**

Debugging SAS® Programs: A Handbook of Tools and Techniques
by **Michele M. Burlew**

support.sas.com/publishing

*Decision Trees for Business Intelligence and Data
Mining: Using SAS® Enterprise Miner™*
by **Barry de Ville**

*Efficiency: Improving the Performance of Your
SAS® Applications*
by **Robert Virgile**

The Essential Guide to SAS® Dates and Times
by **Derek P. Morgan**

*Fixed Effects Regression Methods for Longitudinal
Data Using SAS®*
by **Paul D. Allison**

*Genetic Analysis of Complex Traits
Using SAS®*
by **Arnold M. Saxton**

*A Handbook of Statistical Analyses Using SAS®,
Second Edition*
by **B.S. Everitt**
and **G. Der**

Health Care Data and SAS®
by **Marge Scerbo, Craig Dickstein,**
and **Alan Wilson**

The How-To Book for SAS/GRAPH® Software
by **Thomas Miron**

*In the Know... SAS® Tips and Techniques From Around
the Globe, Second Edition*
by **Phil Mason**

*Instant ODS: Style Templates for the Output
Delivery System*
by **Bernadette Johnson**

*Integrating Results through Meta-Analytic Review Using
SAS® Software*
by **Morgan C. Wang**
and **Brad J. Bushman**

*Introduction to Data Mining Using
SAS® Enterprise Miner™*
by **Patricia B. Cerrito**

*Introduction to Design of Experiments with JMP®
Examples, Third Edition*
by **Jacques Goupy**
and **Lee Creighton**

Learning SAS® by Example: A Programmer's Guide
by **Ron Cody**

The Little SAS® Book: A Primer
by **Lora D. Delwiche**
and **Susan J. Slaughter**

The Little SAS® Book: A Primer, Second Edition
by **Lora D. Delwiche**
and **Susan J. Slaughter**
(updated to include SAS 7 features)

The Little SAS® Book: A Primer, Third Edition
by **Lora D. Delwiche**
and **Susan J. Slaughter**
(updated to include SAS 9.1 features)

The Little SAS® Book for Enterprise Guide® 3.0
by **Susan J. Slaughter**
and **Lora D. Delwiche**

The Little SAS® Book for Enterprise Guide® 4.1
by **Susan J. Slaughter**
and **Lora D. Delwiche**

*Logistic Regression Using the SAS® System:
Theory and Application*
by **Paul D. Allison**

Longitudinal Data and SAS®: A Programmer's Guide
by **Ron Cody**

Maps Made Easy Using SAS®
by **Mike Zdeb**

*Measurement, Analysis, and Control Using JMP®: Quality
Techniques for Manufacturing*
by **Jack E. Reece**

*Multiple Comparisons and Multiple Tests Using
SAS® Text and Workbook Set*
(books in this set also sold separately)
by **Peter H. Westfall, Randall D. Tobias,
Dror Rom, Russell D. Wolfinger,**
and **Yosef Hochberg**

support.sas.com/publishing